CALIFORNIA STATE MINING BUREAU, FERRY BU

BULLETIN No. 57

GOLD DREDGING

IN

CALIFORNIA

LEWIS E. AUBURY

State Mineralogist

SACRAMENTO

W. W. SHANNON, - - - - - - SUPERINTENDENT STATE PRINTING

1910

LETTER OF TRANSMITTAL.

To His Excellency, JAMES N. GILLETT, Governor of the State of California, and the Honorable Board of Trustees of the State Mining Bureau.

GENTLEMEN: I have the honor to transmit to you Bulletin No. 57, "Gold Dredging in California."

This report is the second one on the subject issued under my direction. Owing to the rapid advance made in gold dredging since the issuance of Bulletin No. 36, I deemed it advisable to prepare a report which would furnish information concerning the progress made in this important industry, and also to show the present day methods of operation. In this report will also be found data concerning the dredging lands of the important districts, their value as arable land before dredging, and the primary efforts made towards their use or reclamation after their gold contents had been extracted.

At different times during the past few years, some opposition was encountered towards the industry, and some representations concerning the débris question were made in an effort to effect adverse legislation. While it is admitted that in a few isolated cases there was good ground for complaint, it was unjust that the whole industry should be condemned for the infractions of these few operating companies. In this report I have endeavored to furnish data obtained by the Bureau explanatory of this subject.

The field work on this Bulletin was performed by Mr. W. B. Winston, and Mr. Charles Janin collaborated with Mr. Winston in part of the work of its preparation. I wish to herewith tender the thanks of this Department to both of these gentlemen for their kind and valuable assistance.

I also wish to thank those officers and employees of the different dredging companies who have assisted the Bureau by furnishing valuable information concerning costs, operation, etc., without which the report would be incomplete. Uniform courtesy was extended by the representatives of all dredging companies operating in this State, as well as by the construction companies, to all of whom we wish to extend our kindest acknowledgments.

We wish to also extend our thanks to all others who have in any manner assisted in the matter of furnishing data in the compilation of this work.

It is hoped that with the information furnished in this report, a better understanding of the conditions affecting dredging in this State will be obtained, and that it will serve to still further advance an industry which is of growing importance.

Respectfully submitted.

LEWIS E. AUBURY,
State Mineralogist.

San Francisco, *August 31, 1910.*

TABLE OF CONTENTS.

LIST OF ILLUSTRATIONS.

MAPS.

INDEX TO DREDGING COMPANIES OPERATING IN CALIFORNIA IN 1910.

GOLD DREDGING IN CALIFORNIA.

INTRODUCTORY.

By Lewis E. Aubury, State Mineralogist.

The construction of the first practical gold dredge in California in 1898 marked the beginning of a new era in gold mining, and which branch of the industry has probably, since its inception, assumed greater proportions in this State than elsewhere. With the rapid advances made in gold dredging and the gradual increase of gold output, have likewise come the improvements and enlarged construction, which make the California gold dredge to-day the model after which other countries pattern.

The gold industry of this State has received a new impetus, and the production has advanced approximately three and a half million dollars above the average output of two years ago. Should other forms of gold mining maintain their average two years from now, California will probably again recover its rank as the leading gold producing State in the Union, and will undoubtedly maintain its lead for many years to come. Sufficient area has already been proven in the gold dredging fields to warrant this conclusion. While it has been contended in some quarters that the limits of the dredging areas have already been fairly well defined, and that the large interests and consolidations have a large portion of the fields controlled, new companies are continually being organized for the purpose of exploiting tracts which have been passed over, or others which were prospected a few years ago and "turned down" as being too low grade to operate profitably. With the advent of the modern dredge, handling 250,000 to 300,000 cubic yards of gravel per month, at a cost of from $2\frac{1}{4}$ to 3 cents per cubic yard, conditions have changed materially. The same evolution with dredge mining has occurred as with gold quartz mining in this State, and the question has resolved itself into one of capacity. The present dredge—large as it is—has apparently not yet reached its limits.

A new factor has entered into dredging in California which adds largely to the profits of some of the companies—that is, utilizing the tailings from the dredges. The tailings are broken in immense crushers and the product utilized for macadam for roads and rubble for concrete. What have been termed by some as "the unsightly piles of gravel" have been made to serve various good purposes, and, at the same time, furnish the best class of material at a minimum cost to the consumer.

Several protests have been made in the past against permitting dredge

mining. These protests have been made without an adequate knowledge of actual conditions, it being claimed that the débris from the
dredges was allowed to flow into the river channels to their detriment,
and the whole industry was consequently condemned. Investigation
showed that in a few instances dredging was being carried on in the
streams and some damage was undoubtedly done. Outside of these
instances, however, the balance of the dredges have either operated in
landlocked sections, away from the streams, or were depositing the tailings on the banks of the streams, deepening the channels and improving
them.

Some complaint has also been made of the total destruction of the
soil where dredges operate. Regarding this matter, and as will afterwards be shown in this report, but a comparatively small amount of
arable land is included in that which has been or will be dredged.

On the other hand, reclamation projects are now under way which
are being carried on by some of the dredging companies, which will
restore to cultivation hundreds of acres of swamp or overflowed lands,
and which, were it not for the enterprise of these companies, would
remain waste or unproductive for years to come. These reclaimed lands
will far more then offset those which some claim will be irreparably
ruined. The so-called "destroyed lands," which at one time were productive, had, to a large extent, been made worthless for agriculture,
viticulture, or horticulture by their former owners before a bucket had
turned them over. The mineral ingredient necessary to plant life had
been exhausted from the surface soil, and it was practically impossible
for the farmers to raise a profitable crop from them. To a certain
extent, dredging these lands *has reclaimed them*. Trees or vines planted
since the lands were dredged give ample evidence of the fertility of the
ground, and serve to illustrate the improved nature of the same.

In the dredging sections, lands were purchased by dredging companies
for $25 per acre, and the same land after dredging, with its cobble piles,
was sold for $100 per acre for the gravel contents. After the gravel
has been removed, or the lands leveled, they can again be utilized, if
necessary, for farming lands.

The dredging industry in California, while adding annually millions
of dollars worth of gold to our State's products, at the same time dis-
penses immense sums through the various channels of trade, and adds
materially to our prosperity. It will continue to do so for at least the
next decade, and as actual conditions affecting the industry become
more generally known, and proper restrictions are maintained, it will
be more appreciated, and the objections which have been raised will
disappear.

The possibilities of recovering gold by dredging were not deemed
practical until recent years, and some of the early miners did not believe

its recovery by this means would ever be possible. As indicating some of their ideas, I will quote from a chapter of the Reports upon Mineral Resources of the United States, 1867, by Special Commissioners J. Ross Browne and James W. Taylor:

"It was not, however, in quartz mining alone that ridiculous blunders were made. Large sums of money were expended in the Eastern states by men who had never seen a placer mine, and had no correct idea of the nature of the gold deposits, in making machinery to take gold more expeditiously from the river beds and bars than could be done by hand. One enterprising New York company sent a dredging machine to dig the metal from the bottom of the Yuba River, never questioning whether that stream was deep enough in the summer to float such a machine, or whether the tough clay and gravel in its bed could be dug up by a dredger, and entirely ignorant of the fact that the gold is mostly in the crevices of the bedrock, where the spoon and knife of the skillful and attentive miner would be necessary for cleaning out the richest pockets."

It will be seen from the above that even in those days the idea of recovering gold by dredging was considered chimerical, and while a few adventurous spirits were on the right track, the mechanical means were at fault.

The history of some of the failures attendant on early experiments in dredging are dwelt upon in this Bulletin, as well as in No. 36, so I will not go further into that subject. The present day bucket dredge has apparently not yet attained perfection, but the improvements in construction and operation have been so rapid and so many that little more can be said of the high standard which has been reached in dredge building than to refer the reader to the cost of operations, capacity, etc., of California dredges, which more practically tell the tale.

Outside of the bucket dredge, no others, up to the present time, have been operated in California which have proven successful in recovering gold.

The opportunities for profitable investment in dredge mining in California are many. The field is a large one, and so far as proven extends from Siskiyou County in the northern part of the State to Merced County on the south. Descriptions of this territory will be found in the following chapters. A general description of dredging fields and operations outside of California has also been added to this report. It may prove interesting to the reader as a comparison with conditions found in California.

I believe it can be safely asserted that no better dredging fields can be found than in this State, where conditions as to transportation, climate, power, character of material to be handled, etc., are nearly perfect, and good gold values are also found.

In the following chapters it has been the intention to furnish data of a nature which would meet the requirements of those interested in the subject, and I trust that the information contained in the Bulletin will prove adequate in this respect.

The gold dredging industry in California is established on a stable basis. This is made obvious by the detailed figures that appear in this publication. Its continuance is a reason for congratulation. Much outside capital has been attracted to California by reason of it and the success attending investments has stimulated interest in this State, with results that are beneficial to all classes of citizens. The resources of this State are practically exhaustless. There may be an end to gold discoveries and to the invention of means to reach the precious metals in California, but that date is assuredly far off. For many years, in addition to wealth to be derived from many other sources, we shall hold out to the enterprising and intelligent miner the lure of gold production, and in that consideration gold dredging will have a prominent place.

1. GENERAL, HISTORICAL, AND GEOLOGICAL.

By W. B. WINSTON and CHARLES JANIN.

The history of gold dredging in California practically begins with the floating of the first successful bucket elevator dredge at Oroville, on March 1, 1898, and the rapid rise of the gold dredging industry in California is, in a great measure, due to the enterprise and the successful operations of W. P. Hammon and the late Thomas Couch. These gentlemen contracted with the Risdon Iron Works, of San Francisco, for the construction of this dredge, which operated for several years at the lower end of the Oroville district, on ground now belonging to the Natomas Consolidated of California.

No. 1. First successful dredge in California. Feather River No. 1, Risdon type.

The above dredge, first known as Couch No. 1, and later as Feather River No. 1, was by no means an unqualified success at first. Couch and Hammon experienced many weeks and months of anxiety, and expended large sums of money in changes and repairs before demonstrating that the venture was not a failure. It may be said that during this time the fate of the dredging industry hung in a balance, at least as far as the Oroville district was concerned. It would, undoubtedly, have been many years before other parties would have undertaken experiments, especially as at that time gold dredging was little known of in America, and mining men, in general, had little faith in such ventures, and could not have been relied upon for assistance.

Thomas Couch lived to see his ventures a far greater success than anticipated, and W. P. Hammon, who, since the beginning, has been the leading gold dredging operator in California, continues to-day in control of the operations of the largest number of dredges and the largest gold dredging companies in America.

It required many years of evolution for the gold dredging industry to reach the present stage of practical success. Among the first ideas that occurred to many of the early gold hunters in California was the use of a machine to scoop up the gravel from the inaccessible beds and bars of auriferous streams, and it was only a few months after Marshall's discovery of gold in California that a machine was shipped around the Horn from New York to San Francisco. It arrived in 1849, and was soon at the bottom of the Sacramento River. During the succeeding years many attempts at gold dredging were made in California, and elsewhere in the Western States of America, but with the exception of the operations with double-lift-bucket elevator dredges of the Bucyrus type, at Grasshopper Creek, Montana, in 1894, all proved failures.

It was not until 1897 that a dredge of the single-lift-bucket elevator type was floated in California. This dredge, which was constructed by the Risdon Iron Works for R. H. Postlethwaite and floated on the Yuba River, would probably have been a success if located at Oroville; being, however, in a turbulent stream, where afterward it was found difficult to even hold piling during the flood season, the dredge was wrecked and was not recommissioned.

Little advantage would result from narrating the history, in detail, of the early failures in gold dredging, or from giving a detailed description of the obsolete machines designed for this purpose. Dredging has become an established and important branch of mining, and, with the information at hand, the early failures are not needed to warn careful investors against similar mistakes; it is sufficient to say that among the machines that have not proven a success so far in gold dredging are the submarine boats, the pneumatic caisson, the hydraulic or centrifugal pump suction dredge, the clam-shell, and the vacuum dredge.

The steam shovel, or steam paddy as it was originally called, was used in America for excavating purposes many years prior to gold dredging. This and the original single bucket, or spoon dredge, as evolved in New Zealand in the early sixties, were probably the forerunners of the present placer mining dipper dredge, which in several instances has proven a success in California. The single bucket or spoon dredges of New Zealand were at first crude machines, consisting, in a general way, of a bag laced or riveted to a round iron frame secured at the end of a long pole, and drawn along the bottom of the creeks or rivers. This bag was so counterbalanced that when filled, or partly so, it could be hauled up and the contents emptied into a sluice box or rocker. The next improvement was the use of pontoons in connection with the spoon, an auxiliary scow being used for the washing apparatus. These dredges were at first operated entirely by hand, and afterward by current wheels, the first steam-driven dipper dredge, as invented by Ward, being built about 1870.

The first experiments with placer mining bucket elevator dredges were made in 1867 at Otago, New Zealand. These dredges were operated with power furnished by current wheels, the first steam driven bucket elevator dredge being constructed to operate on the Molyneux River in 1881.

History relates that a bucket elevator dredge, built about 1882, operated successfully for sixteen years, and on account of the success of a dredge of this type in 1889, on a branch of the Molyneux River in New

No. 2. Model of native dredge in Ambos, Camarines, Luzon, Philippine Islands.

Zealand, some twenty were built at an average cost of $17,500. Their operations, at first, however, proved failures, probably owing to unsuitable conditions of the gravel deposit, and perhaps also to inexperienced or incompetent management. Many of these dredges were later floated down the Molyneux River, and, under new ownership, proved successful. A number of suction dredges were built in New Zealand, but, as elsewhere, were failures.

The reason for some of the early failures in California is indicated in the illustration entitled "An Old Timer." This machine, on which about $40,000 was expended, was equipped with a bucket ladder made of two pieces of 4-inch by 18-inch Oregon pine. The buckets were made of No. 14 iron and the links of ⅜-inch by 2-inch tire-iron, and bolted together with ½-inch carriage bolts. The machine was at first driven by steam and afterward by a gasoline engine, and is reported to have handled 8,000 cubic yards of gravel during a period of about two years and three months, while operating in ground that is said to have yielded 30 cents per cubic yard.

Although gold dredging had been successfully carried on in New Zealand for many years prior to this mode of mining in California, it is doubtful if the idea of gold dredging in California emanated entirely

No. 3. "An old timer." Illustration of one of the early mistakes in dredge construction.

from the success met with in New Zealand. Among the first men to be consulted by W. P. Hammon were F. T. Sutherland and W. H. Christie. Sutherland, who knew little about dredging in New Zealand, was familiar, however, with the successful operations of the double-lift-bucket elevator dredges at Bannock, Montana, and interested Captain Thomas Couch, of Butte, Montana, in Hammon's ventures. At the same time Sutherland was trying to convince Captain Couch of the possibilities of dredging the Oroville gravels, W. P. Hammon consulted with W. H. Christie, who, through the firm of Christie & Lowe, sent a Mr. Brown to make an examination of the ground, and to report upon the practicability of its being worked by dredges. The firm of Christie & Lowe, general contractors, were at that time engaged in canal and jetty work in the East, and based their opinion upon their experience in the handling of gravel and sand for economic purposes. Mr. Brown reported favorably, and soon afterwards Captain Couch contracted with the Risdon Iron Works for the building of the first single-lift-bucket elevator dredge in the Oroville district.

The bucket elevator is the only type of dredge that has been a financial success in gold dredging, on a large scale, either in California or elsewhere. From 1898 to 1902 the principal dredges in California were of the single-lift-open-link-bucket elevator type, equipped with tailing stacker; in 1899 and 1900, two double-lift-open-link-bucket elevator dredges, equipped with tail sluices and tail scow, were constructed, one by F. W. Griffin and D. P. Cameron and one by the Bucyrus Company, and put in operation at Oroville and at Folsom, respectively, the Continental and Ashburton No. 1. It was not until 1 9 0 1 that dredge constructors abandoned the double-lift dredges, and centered their attention to the perfecting of the single-lift type, equipped with close-connected buckets and belt tailing stacker and driven by electricity in place of steam. It may,

No. 4. Old type of dredge, showing tail sluice and tail scow. Continental Dredge, in 1899, before remodeling.

therefore, be said that the history of the large modern California gold dredge commenced with the construction of the first electrically driven single-lift-close-connected-bucket elevator dredge in the Oroville field in 1901. The first dredges in the Oroville district were equipped with 3¼-cubic-foot buckets; on some of the new dredges these were increased, in 1899, to 5-cubic-foot, but the majority of the dredges up to 1904 were of

No. 5.   Hauling dredge machinery in the mountains, Siskiyou County, California.

3-, 3¼-, and 4-cubic-foot size. The Marion Steam Shovel Company, in October, 1904, constructed a 7-cubic-foot-bucket elevator dredge for the Boston and California Gold Dredging Company. At the present time there are five 7-cubic-foot and three 7½-cubic-foot dredges in the Oroville district, the others being 3-, 3½-, 4-, 5-, 5¾-, and 6-cubic-foot capacities.

While, since the beginning of dredging operations in California, the Oroville district has been the largest gold producer from this source in

No. 6.   Hauling dredge machinery in the mountains, section of digging ladder.

the State, and has had the greatest number of dredging companies and dredges, the advancement in dredge construction and in the perfection of the large dredges of to-day was more rapid in the Folsom or American River district, where as early as 1899 the first 7½-cubic-foot dredge in California was constructed by the Bucyrus Company for R. G. Hanford. This was put in operation by the New England Exploration Company, March 1, 1900. In 1905 the first 8¼-, 9-, and 13-cubic-foot dredges in California were put in commission along the American River in the Folsom district.

The principal dredging fields in California, up to 1903, were located along the Feather River, near Oroville, and on the American River near Folsom. Since 1903, dredging along the Yuba River between Marysville and Smartsville has materially added to the gold output of the State, and, owing to the depth necessary to dig, from 50 to 75 feet, many new features in dredge construction have been developed. This district is now one of the largest gold dredging fields in the world. The land is owned by two companies, one of which has twelve dredges operating and one under construction; and the other, an operating plant of three dredges.

Aside from the Oroville, Folsom, and Yuba districts, there are three dredges operating in Calaveras County, one in Merced, one in Shasta, two in Siskiyou and one in Stanislaus, making a total of 63 operating in the State, which, together with three large dredges under construction, represent an investment of about $7,205,000.

Owing to their small capacity and to the great improvement in dredge construction, there are 38 dredges, representing an investment of about $1,790,000, now idle or dismantled. The following table gives the number of elevator dredges commissioned in California from 1897 to 1910, as nearly as can be ascertained, the number in active operation, and approximately the capital invested in them:

District.	Dismantled	Operating in 1910	Constructing in 1910	Total	Capital invested†
Butte County—					
Oroville	21*	25	1	47	$3,135,000
Wyman's Ravine	1*	2		3	⎫
Honcut Creek		2		2	⎬ 375,000
Butte Creek		1		1	⎭
Sacramento County	6	9	1	16	2,010,000
Yuba County	1	15	1	17	2,220,000
Calaveras County		3		3	⎫
Merced County		1		1	⎪
Stanislaus County		1		1	⎪
Placer County	4	1		5	⎬ 1,255,000
Shasta County	2	1		3	⎪
Siskiyou County	1	2		3	⎪
Trinity County	2			2	⎭
Totals	38	63	3†	104	$8,995,000‡

* This includes four dipper dredges which operated successfully in the Oroville district.
† All of these dredges are to be equipped with 13½-cubic-foot buckets.
‡ Approximated.

There is, perhaps, no branch of mining which has come to the front in so short a time as gold dredging, which from 1898 to 1908 produced over $25,000,000 worth of placer gold in California. On account of hydraulic mining being prohibited in many counties, the output of placer gold in California would have shown a marked decrease, in late years, had it not been for dredge mining.

The following diagram shows the rapid increase in the gold output from dredging operations:

Production of Gold Won from Dredging Operations in California, from 1898 to 1909.

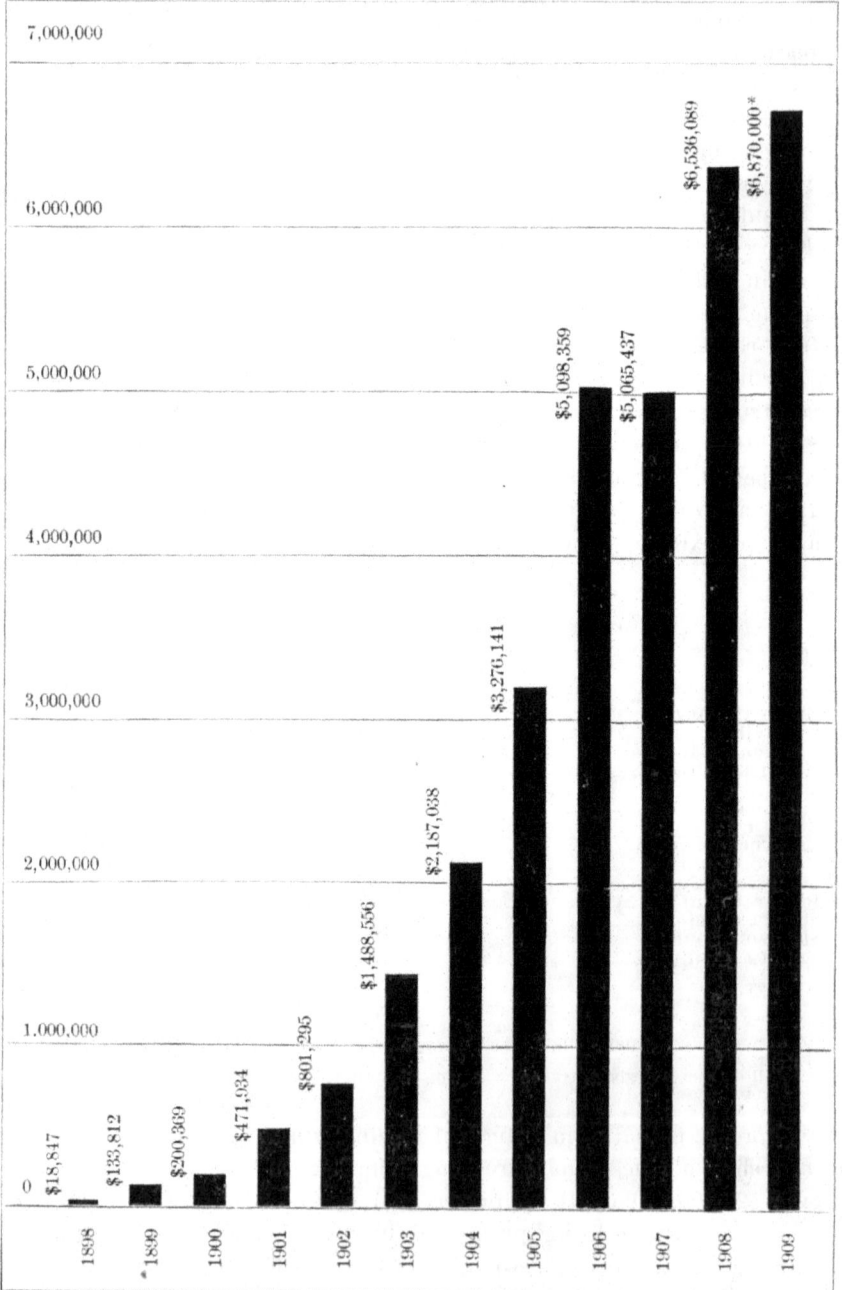

* 1909 estimated.

The different methods of mining the gravel deposits of California are as follows: first, river, bar, and gulch mining, or washing the detritus, either in or closely adjacent to the beds of the present rivers; second, hydraulic mining; third, tunnel or drift mining, which is working the rich portions on the bedrock by means of drifts; fourth, what has been called, for want of a better name, surface mining, which is intermediate in character between the river and hydraulic mining processes; and fifth, dredging, the most recent method of working the gravel deposits, where natural conditions allow the installation of a plant.

In 1903 the gold production from dredges exceeded that from either the hydraulic or drift mines. In 1904 the yield exceeded that of both the hydraulic and drift mines combined, and in 1905 the dredges produced $461,205 more than the drift, hydraulic, and surface placers together, and since then the output from dredging has exceeded the total yield of the three other forms of placer mining. In 1908, the dredges produced $4,841,191 more gold than all the hydraulic, drift, and surface placers together, or over 79 per cent of all the placer gold of the State for that year. In 1908 there were eight counties reporting returns from dredge operations. The following table shows the production of gold from dredging operations in California from 1898 to 1908, by counties:

Production of Gold Won from Dredging Operations in California, from 1898 to 1908, Compiled from U. S. Mint and Geological Survey Records.

Counties.	1898.	1899.	1900.	1901.	1902.	1903.
Butte	$18,847	$132,412	$154,065	$396,919	$614,380	$1,329,998
Calaveras					3,948	12,807
Madera, Merced, and Stanislaus						
Sacramento			17,200	47,619	155,194	102,097
Shasta						
Siskiyou		1,400		5,004	18,773	7,318
Trinity			29,104	22,392	9,000	10,600
Yuba						25,736
El Dorado						
Totals	$18,847	$133,812	$200,369	$471,934	$801,295	$1,488,556

Counties.	1904.	1905.	1906.	1907.	1908.	Grand Totals.
Butte	$1,632,507	$2,261,887	$2,768,782	$2,697,092	$3,043,051	$15,049,940
Calaveras	115,951	202,505	177,112	*30,802	198,600	741,725
Madera, Merced, and Stanislaus					182,970	182,970
Sacramento	348,990	569,124	921,300	649,511	1,109,196	3,920,231
Shasta					30,966	30,966
Siskiyou	6,827	7,111	26,000		2,227	74,660
Trinity	8,500	5,000				84,596
Yuba	74,263	188,967	1,205,165	1,688,032	1,969,079	5,151,242
El Dorado		41,547				41,547
Totals	$2,187,038	$3,276,141	$5,098,359	$5,065,437	$6,536,089	$25,277,877

* Shasta County included.

In 1908 the total gold production amounted to $18,761,589, in California, of this $10,530,372 came from deep mines, $6,536,189 from gold dredging, and $1,694,998 from hydraulic mines and surface placers.

Of the gold producing counties in California in 1908, there were four which yielded no placer gold, three which yielded no quartz gold, and seven, the production from which was over $1,000,000 each; the rank in value of output of these counties is as follows:

Butte	$3,139,398	Calaveras	$1,378,511
Nevada	2,297,963	Sacramento	1,166,055
Yuba	2,034,486	Shasta	1,131,832
Amador	1,876,174		

Butte, Sacramento, and Yuba are dredge mining counties, the others being principally quartz mining. Butte ranks the first gold producing county in the State, which is due to the large number of gold dredges operated there. Sacramento showed the largest increased yield of gold for the year 1908, $375,082, this being due to the operations of new dredges as mentioned elsewhere. The largest production of gold from deep mines came from Nevada County.

The following table shows the number of dredges operating in California, the gold produced and the number of persons employed during 1908 and 1909:

Year.	Dredges Working.	Put in Operation During the Year.	Gross Yield.	Yield per Dredge.	Number of Men Ordinarily Employed.
1908	65	9	$6,536,089	$100,555.21	940
1909	64	5	6,870,000	107,328.12	932

Several of the working dredges were constructed and put in commission during the year, while others were either dismantled or delayed by minor repairs, so that few of the dredges operated continuously; while the number of operating dredges are apparently on the decrease, the gold production from this source is increasing, and it is expected

No. 7.   Hauling dredge machinery in the California mountains.

that the output in 1910 will exceed $7,000,000. The dredges now constructed are usually equipped with buckets from 7½ to 13½-cubic-foot capacity, and it is probably only a matter of time until 15-cubic-foot buckets will be used. The early dredges had a capacity of from 25,000 to 45,000 cubic yards of gravel per month, while the large dredges of to-day handle an average of from 160,000 to 250,000 cubic yards per month.

Owing to the financial success of the gold dredging industry, most of the gravel deposits in the State have been explored. It is doubtful if any new dredging fields, as rich as those now being worked, will be found, but it is possible that with the increased capacity of the dredges, many low grade deposits will be worked in the future.

The present known payable districts in California which have so far been proven comprise an area of about 19,000 acres, and are located in Butte, Yuba, Placer, Sacramento, Calaveras, Stanislaus, Merced, Shasta, and Siskiyou counties. The following table shows, in a general way, the extent of the dredging ground in these counties, the average depth of the gravel, and the value per cubic yard. Much of this ground has already been dredged:

Counties.	Total Proven Dredging Ground.	Average Value per Cubic Yard.	Average Depth of Ground.
Butte	6,600 acres	15 cents	30 feet
Yuba	3,600 acres	15 cents	65 feet
Placer	430 acres	9 cents	38 feet
Sacramento	6,050 acres	12 cents	35 feet
Calaveras	850 acres	15 cents	18 feet
Stanislaus	200 acres	15 cents	22 feet
Merced	400 acres	14 cents	20 feet
Shasta	700 acres	12 cents	22 feet
Siskiyou	370 acres	14 cents	35 feet

A number of years ago two Risdon dredges operated in Trinity County, near Trinity Center and at Poker Bar; there is now, however, no gold dredging being carried on in the county. Several tracts are now being considered on which to install new dredges. In El Dorado County some dredging operations were started a number of years ago, but are not reported to have been a success and no dredging operations of importance are being carried on in the county at the present time. A new bucket dredge is being constructed at Cache Rock, in Placer County, by the Risdon Iron Works. At this point a Dubois suction dredge was erected a few years ago, but it is reported that after being in operation for six months only one half ounce of gold was recovered. There are no dredges operating in Plumas County, although considerable prospecting was done a number of years ago. In Shasta County there is at present only one dredge in active operation. The total dredging area is estimated at about 6,000 acres, of which nearly 1,000 acres have so far been proven; considerable prospecting is being done in this county.

Occasional complaints have been made during the last few years, by men not familiar with conditions in the field, about dredges obstructing the rivers and discoloring the water in the streams; and a great deal has been said at times about the destruction to orchards and vineyards by dredge mining. Investigation, however, proved that these objections were unwarranted.

No. 8. View across Feather River, below Oroville, Cal.

Practically all of the dredges now operating in California are working inland and not in the river beds proper; and are not, therefore, obstructing the rivers with tailing, causing overflow during flood times, or damage to adjoining lands. The dredge men have for years been impounding the overflow from the dredge ponds and none of this water is allowed to flow into the rivers.

While good orchard land has been destroyed by dredge mining, this is more than offset by the benefit derived by the people of the various communities where dredge mining is carried on, and good results have been obtained in replanting dredged land. The number of profitable orchards destroyed is very small, probably not exceeding 1,000 acres in

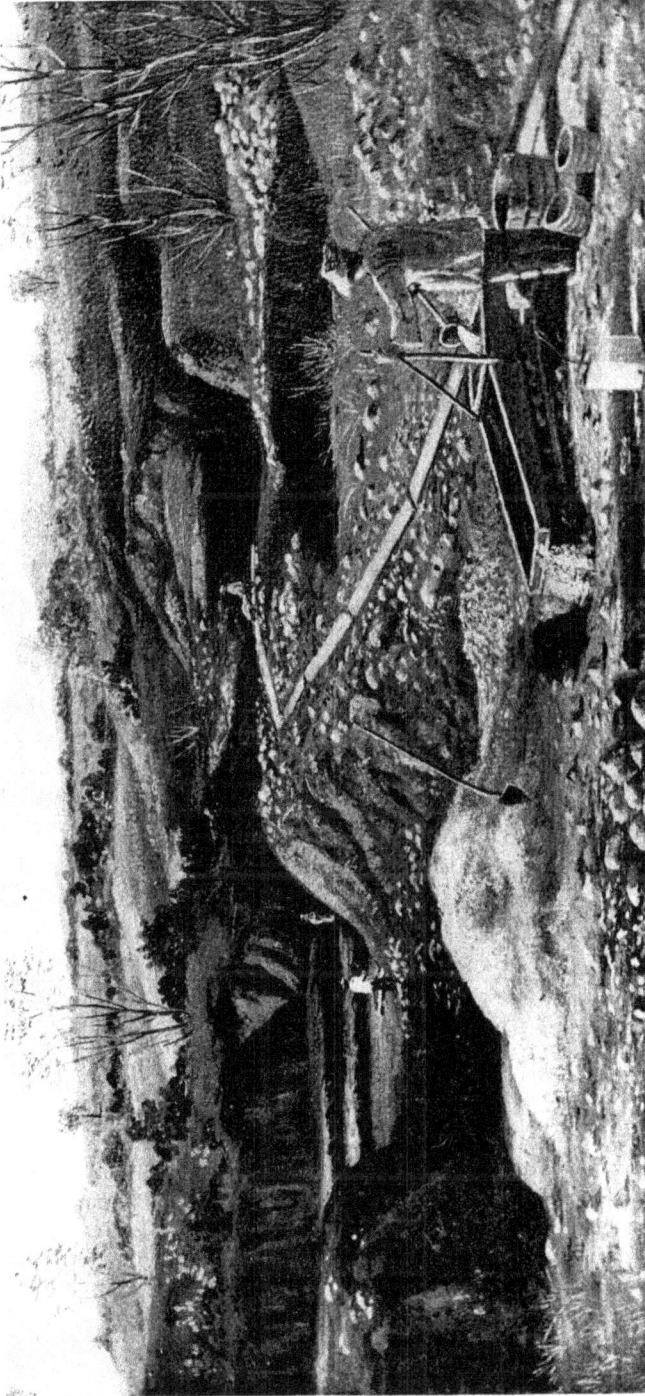

No. 9. Showing method of working ground below Oroville in 1860 and '70. From an old oil painting.

the State, and most of the land being dredged was originally unsuitable for either horticulture or agriculture, and had nearly all of it been mined by the different methods in use previous to gold dredging.

The following table shows the number of acres of land owned by dredging companies in the different counties, the number of men employed in connection with dredging operations and the wages paid them:

Counties.	Acres Owned by Dredging Companies.	Number of Men Ordinarily Employed.	Wages Paid Yearly. Estimated.
Butte	8,900	550	$575,000
Yuba	5,500	450	525,000
Placer	990	12	14,400
Sacramento	12,500	340	338,000
Calaveras	850	40	⎫
Stanislaus	1,000	14	⎮
Merced	500	12	⎬ 134,600
Shasta	2,000	15	⎮
Siskiyou	450	35	⎭
Totals	32,690	1,468	$1,587,000

It may be of interest to note that the relation of this branch of mining in California to neighboring interests has in every instance been such that wherever dredge mining is carried on, the ranches and towns have been directly benefited. To-day great activity and prosperity exist in Oroville, Marysville, Hammonton, Marigold, Folsom, Jenny Lind, Callahan, etc., places which were almost dormant, or did not exist previous to the advent of dredging companies.

In addition to gold mining, work of considerable magnitude and economic importance is also being carried on by gold dredges. On the Yuba River construction of gravel embankments to confine and control the flow of the Yuba River in the vicinity of Daguerre Point, is being done for the United States and the State of California, under agreement with W. P. Hammon and associates in exchange for the right to extract gold from the gravel used in making the embankments. These embankments are to be about 20,500 feet long, 300 feet wide on the base, and 30 feet in height at the highest places.

GEOLOGICAL.

A large portion of the gold mined in the world comes from the detritus of gold-bearing veins and rocks caused by eroding and disintegrating agencies which have been at work ever since rocks were formed. While there are localities where gold has been recovered in large quantities from placers where no vein mining has been successful or important quartz veins discovered, careful investigations would probably prove that the gold in the detritus came from auriferous seams and veins and from the rocks immediately adjacent to them.

It is an interesting fact that the largest dredging fields in California are located on three of the large rivers which drain the principal mineral region of the State, on the western slope of the Sierra Nevada, and where these rivers, leaving their narrow canyons, break from the foothills into the Sacramento Valley. The region drained by the different branches of these rivers, which are the American, Yuba, and Feather, is about 350 miles long, from 30 to 70 miles wide, and has an elevation of from 500 to 5,000 feet. It includes within its boundaries a great portion of the ancient river channels which were mined with considerable profit in early days, and the present river systems, while draining practically the same area as that drained by the ancient rivers,

No. 10.   General view of the Folsom dredging ground, showing old placer pits previous to dredging.

run at right angles to and cut across the ancient channels in numerous places. The placer deposits of these present streams are for the most part the river banks, bars, benches, and flood plains. The gold they contain came from the disintegration of the auriferous slates and other rocks and the quartz lodes of the bedrock series, and also from the ancient channels that were cut and eroded by the present rivers.

In some districts portions of the ancient channels can be traced for long distances, and their continuity and identity has been established with certainty by the mining operations; in other districts the remains of the old channels are fragmentary. At Oroville, in Butte County, is the debouchure of a great river coming from the north, corresponding to the present Feather River and draining the same territory. At Smartsville, in Yuba County, is the evidence of an ancient river, apparently the counterpart of the present Yuba. In Placer County, there are remains of an old river channel, the predecessor of the American. At La Grange, in Stanislaus County, is the outlet of the early rivers of Tuolumne and probably Calaveras and Amador counties.

The filling of the ancient channels is composed of gravels of white quartz, metamorphic schists, etc., with layers of light colored clays and sands, and they are generally capped with volcanic flows. Certain layers of the gravels are cemented, probably owing to percolation of silicious and calcareous waters. The present river gravels consist of well rounded pebbles of silicious and volcanic rock and quartz. The size of these pebbles ranges from a fraction of an inch to cobble stones six or eight inches in diameter, though the average size is less than these latter dimensions.

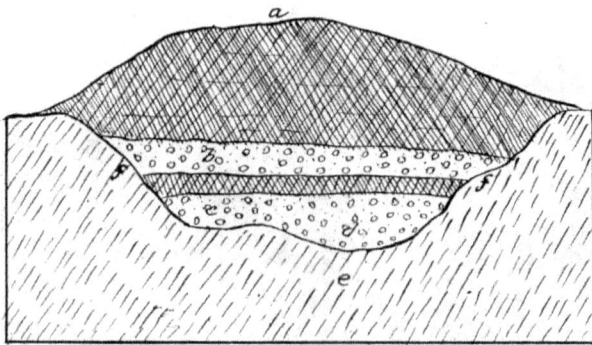

No. 11.   Ideal section old gravel channel after R. E. Brown.
a, volcanic cap; b, upper lead; c, bench gravel; d, channel gravel; e, bedrock; f, rim.

On the bedrock, large partly rounded fragments occasionally occur, and sometimes well rounded boulders of granite and other rocks several feet in diameter are found. All the gold of the valley placers is relatively fine as compared to the coarse gold found in the higher altitudes of the present rivers, and the finest particles may have been carried a great distance; the coarse gold has probably not traveled far from where it first lodged. Undoubtedly much of the finest gold recovered in placer operations was first associated with such sulphides as pyrites, galena and arsenopyrites, and was freed from this association by the decomposition of these sulphides and left behind as free gold. The smoothing of gold grains is caused by impact with the sand and gravel sweeping with and over them.

The velocity of the water bringing down the detritus from the mountains regulates the distribution, and, to a certain extent, the concentration of the heavy minerals. Where the gravel deposit is loose or porous, permitting the circulation of water, mechanical concentration of materials after the gravel is brought down from the mountains still continues in the valleys, caused either by running or percolating waters shifting the gravel and allowing the gold and other heavy minerals, such as black sand, to sink naturally toward the bottom, so that the greatest percentage of the gold found in the gravel is generally found in the lower portion of the deposit and sometimes within a few inches of the bedrock.

At times the gold occurs more evenly distributed throughout the whole deposit, caused, probably, by a more regular flow of the distribut-

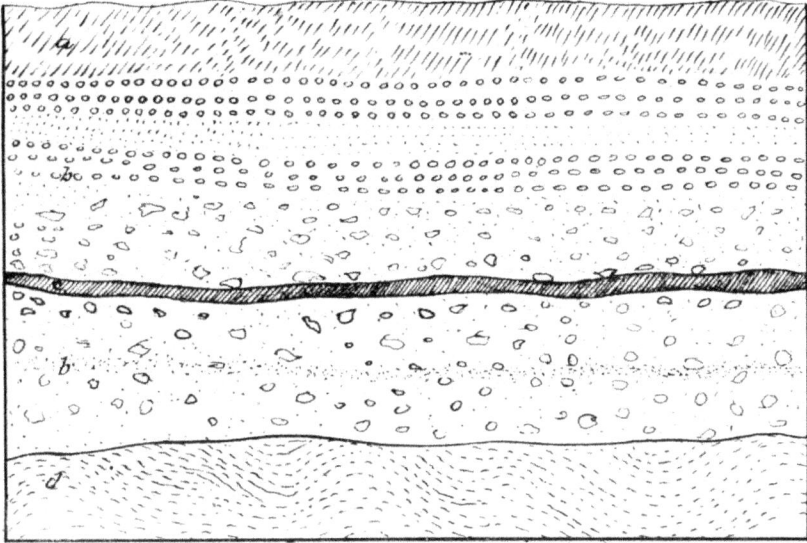

No. 12. False bedrock in placer deposit.
a, top soil; b, gravel; c, false bedrock; d, true bedrock.

ing stream and slight after concentration. Where occasional seams of clay or other material form an impervious layer in otherwise porous gravel, the downward movement of the heavy minerals is arrested and above these layers "pay-streaks" will be formed, sometimes several of

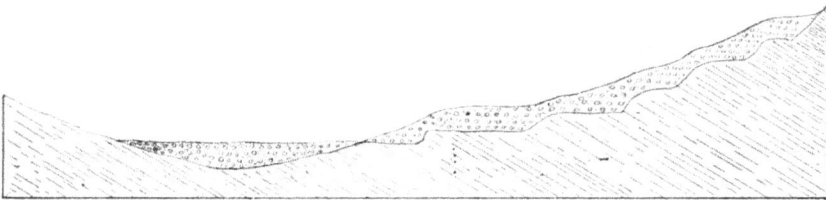

No. 13. Section showing bench and valley placers.

these "pay-streaks" being shown in a cross-section of a deposit. The gravel in Sacramento Valley district lies, in general, on a stratum of volcanic ash, forming a false bottom or bedrock. This bedrock overlies other gravel, the true bedrock being the slates and schists which form the adjacent foothills. In the northern counties, like Siskiyou, Trinity, and Shasta, the deep canyons of the rivers, especially the Klamath and its branches, contain large amounts of gravel which lie upon a true bedrock. When the bedrock is a slate shale or schist, affording cracks for the gold to settle in. it is often found profitable to work it for several inches in depth.

There are often several benches or terraces along the sides of a river which were a part of the river bottom when the stream was at a higher level. The dredges on the American and Feather rivers work at different elevations on the different benches, the former channels of the present streams. On the Yuba River, owing to the great quantity of

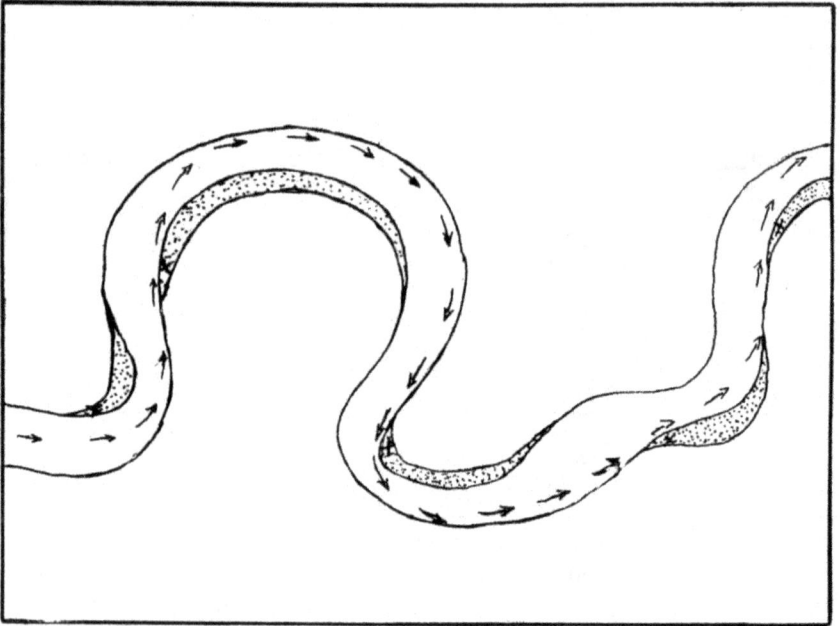

No. 14.   Ideal River, showing accumulation of bars.   Crosses show the most favorable spots for the deposition of gold.   After J. B. Spur.

hydraulic tailings covering the river valley, the dredges work at a more uniform elevation.

In washing gravel concentrates of heavy minerals, other than gold, are also collected; magnetite becoming the miner's black or magnetic sand, and garnets, which are found in schists and other metamorphic rocks, the ruby sand; in addition to these platinum is sometimes found in appreciable quantities.

The presence of black and ruby sand is regarded most favorably in prospecting, and indicates that there has been a concentration of materials in the gravel and the possibility of gold being present. though it does not necessarily mean that gold will be found.

II. PROSPECTING DREDGING GROUND.

To make a gold dredging proposition an economic success, experienced judgment must be used in every phase of the work; preëminently is this necessary in the details of prospecting or examination.

There have been many failures in dredging because a thorough examination of the ground was not made before installing a dredge. The tendency to build an expensive plant before the value of the property has been sufficiently demonstrated is as pronounced a fault in the dredging industry as in other branches of mining.

The development and improvements in the construction of machinery, and the overcoming of what at first seemed insurmountable difficulties, have greatly broadened the dredging field and relieve the engineer of to-day of many of the difficulties that he was confronted with a few years ago. The thorough investigation of a placer deposit, however, is not so simple a matter as the uninitiated investor is often led to believe; such investigations should never be carried no by one unfamiliar with the important factors governing a proper determination of the value of placer ground, and with the conditions necessary for the successful operation of dredges.

It is difficult to give any one factor as being the most important in the examination of dredging ground, as several conditions must be favorable to insure success. The gold value in itself may be of minor importance in the valuation of some ground, as a hard uneven bedrock, the presence of many large boulders, a great quantity of clay, or a rough surface contour might prevent the economic dredging of ground with a comparatively high gold content; it may be said, however, that sufficient gold content, the feasibility of economic operation, and the presence of enough ground to warrant the cost of equipping at least one dredge, are the salient features.

To obtain information concerning the value and character of the gravel and of bedrock it is necessary to sink either shafts or drill holes; frequently gulches, old prospect shafts, pits, or hydraulic faces, are available and give considerable information, in a reconnaissance of the ground, perhaps sufficient to warrant an engineer of experience in advising his clients to go to no further expense in the examination of the property; he would not, however, be justified in reporting favorably upon the proposition without making a complete examination. In addition to those conditions which influence the actual operation of a dredge, such as depth and character of gravel, and character of bed-rock, fuel or power possibilities, labor conditions, etc., there are others of almost equal importance, as their unfavorable determination would result in the turning down of a proposition that at first glance seemed meritorious; among these are climatic conditions, transportation facilities governing the cost of supplies and installation, price of ground, titles, etc.

Proper examination of placer ground in California for dredging, therefore, involves the determination of the following conditions:

1. Value, character, and distribution of gold content.
2. Depth, character, and quantity of ground to be worked.
3. Character and contour of bedrock.
4. Water level and available supply of water.
5. Costs of fuel, power obtainable, possibilities for hydro electric power.
6. Labor, transportation, supplies, etc.
7. Surface contour and timber growth.
8. Operating costs.
9. Cost of land, royalties, titles, etc.
10. Elsewhere in America the climate might have an important bearing, and in foreign countries other conditions to be considered would be the climate, government, taxes, duties, etc.

In deciding upon a method of prospecting, shafts should always be sunk, if possible, in preference to drill holes, as the results from drill tests are less to be depended upon in regard to the physical conditions and the probable gold content of the gravel than those from shafts. Shaft sinking enables a much larger sample to be taken, gives a better opportunity of examining the character of the gravel, and if the material taken out is carefully handled, there is less chance for error in the work than in drilling. Shaft sinking, however, is limited to favorable conditions while drill tests can be made in any class of ground, and it must be remembered that the value of nearly all the dredging ground in California has been computed from sampling by means of drills.

The chances for making errors in prospecting are great, and the operation is one that requires constant care. In drilling, a careless runner may neglect to drive the pipe ahead of the drill, and pump out an excessive quantity of material when pumping the drillings from below the driving shoe, as has frequently been done. This may sometimes account for the indication of high values that are not confirmed in subsequent working of the ground. In loose ground, the pipe may be driven too great a distance before pumping and the proper amount of material not secured, as the drill pipe may become clogged and the core fail to increase in proportion to the depth the pipe is driven. The element of risk from salting in securing samples from either shafts or drills depends upon the conditions of examination, care taken, and the experience of the engineer.

When it is impracticable to sink shafts on account of the presence of considerable water, prospecting must be done with drills and experienced men should be engaged, and every precaution taken to insure that the results obtained are as nearly as possible indicative of the gold content of the ground prospected.

The gravel may be fine and sandy, medium or coarse. Fine gravel

No. 15.   Chinaman rocking gravel from shaft.   Oroville District.

is one in which pebbles are usually under two inches in diameter, and generally much smaller; in medium gravel, the pebbles run in size up to six inches in diameter; and in coarse gravel much of the material is larger than six inches.

Compact gravel, if not cemented, can be worked with a pick, and a pipe can be driven in same without much difficulty. Loose gravel is easily worked with a pick, and is one that would require lagging in shaft sinking. Quicksand or fine loose gravel containing an excess of water is liable to give an excess of material when drilling for the distance sunk, and the results from shaft sinking in such ground, would probably also be too high, if indeed it was possible to sink shafts.

In sinking prospecting shafts, it is best to use a uniform diameter from top to bottom, as it facilitates computation, if the entire content of the shaft is to be washed, but it is often the practice to cut out a small section along the side of the shaft from top to bottom instead of washing the total amount of gravel. Square shafts are not often used unless the ground is too loose or wet to stand and needs timbering, in which case shaft sinking becomes expensive and soon passes the limit of economy. In sinking shafts in wet ground, it is also the practice to use round iron caissons, usually in four-foot sections, in place of timber lagging, and to sample the entire content of the shaft. In very wet ground, however, if a shaft is sunk at all, it is best to check it with a drill hole. In case where a heavy flow of water is encountered in shaft sinking, the shaft can be discontinued and the remaining distance to bedrock be sunk with a drill hole.

A convenient size for round shafts to a depth of 30 feet is 36 inches, and for depths of from 50 to 60 feet, 40 inches diameter. While shaft sinking is generally considered cheaper than drill tests, the cost varies a great deal with different localities and conditions. Under favorable conditions, the cost of shaft sinking ranges from 50 cents to $2 per foot, according to local wages. Timbered shafts in prospecting gravel have cost as much as $25 per foot.

In prospecting with drills in California, the Keystone No. 3 traction machine is generally used. This is a self-contained machine, equipped with an 8- or 10-horsepower boiler, and operates the drill by means of a walking beam. For fuel, wood, coal, or oil can be used, or where electric power is conveniently available, the boiler can be discarded for an electric motor.

The casing generally used is about 6 inches inside measurement, 5-16-inch thick, and weighs about 28 pounds per foot, and is cut in sections of from 5 to 7 feet long, which are added as additional pipe is required. It is necessary to have a number of extra lengths on hand as it is at times impossible to withdraw all the casing from deep holes. The outside diameter of the cutting shoe is about 7½ inches. The threads of the pipe and couplings are cleaned from grit, and are generally slightly lubricated with axle grease, care being taken not to allow any grease to get in the hole. It is important to see that the threads of the pipes and couplings are in good condition before using, to prevent the loss of the pipe in the hole when pulling.

Drilling without casing has been done in hard ground that will stand without the pipe, but it is dangerous practice, and if the result shows high values, they can not be accepted as truly indicative of the gold content. It may also be the fact that ground hard enough to stand drilling without the casing may be too hard for the dredge to handle.

The casing pipe is marked for each foot in depth, using plain figures, or if the operator desires, some private mark; the drill rope is also marked, records of depth and all details of operation usually being kept by the panner. If the record of pipe depth is not accurately kept, it would be impossible at times to determine whether the drill bit is striking above or below the cutting shoe. For the purpose of a constant base to measure from, it is best to place a fairly long, wide and thick board on the ground near the drill hole, on which the drillman can also stand

No. 16.   Electric-driven Keystone drill.

while working. It is important that the marking of the drill rope be occasionally checked as the ropes stretch and the drill bit wears with use.

The drill bit and stem, weighing from 800 to 1,000 pounds, are allowed to drop with a slight slack in the cable, thus loosening the ground by the impact of its weight on the gravel. Tests made have demonstrated the advantage of quick long strokes from 36 to 40 inches, the drill being arranged to deliver from 55 to 60 strokes per minute. With a slow stroke the loosened material is apt to settle between strokes, causing a recutting of most of the material and a possible loss of gold, aside from a loss of time, as the operation would take longer than if the drill bit struck the clean core each time; the saving of time, however, is a secondary consideration in sampling placer ground, accuracy of results being the prime factor.

There are several types of drilling bits, the one most suitable for gravel and sand being a thin bladed bit; when drilling large boulders or into hard bedrock this bit is sometimes replaced by a heavy rock bit made with a wider angled cutting edge. This second bit, if used in loose gravel, would pack the material, and might cause some to be driven from below the cutting shoe to the side of the pipe. It is important that the drill bit be kept sharp and well beaten out at the edges. Excessive drilling, caused by using a dull bit, may cause a flouring of the gold.

The sand pump is a vacuum pump made of a hollow steel cylinder, 8 feet long and 4 inches in diameter, equipped with a valve on a piston or sucker rod which travels the whole length of the cylinder, going to the bottom of the pump when lowered, and when drawn up rapidly produces a vacuum which draws in sand, slushy water, and small stones. The plunger must fit closely, as the efficiency of the pump depends upon the suction caused by the rod being lifted quickly. The valves of the pump occasionally require repacking, and sometimes a leakage is caused by an obstruction to the valve seat, which is readily repaired. A good pump will draw up everything in the casing that is loose and small enough to be taken in, gold and other minerals are drawn in with the sludge, and are held in the pump by the foot valve in the shoe. In drilling above the water level or in ground containing little water, some water must be kept in the casing, both to facilitate drilling and permit pumping. There are various mechanical troubles that occur during drilling operations that are not worth enumerating and are easily repaired by an experienced drill runner.

In starting a drill hole from the surface, a shallow hole, similar to a post hole, is usually dug and the first section of the casing to which the digging shoe is attached is tamped firmly in and plumbed with a spirit level. Care must be taken to keep the pipe constantly perpendicular. for if this is not done, the casing may become bent, making it necessary to abandon the hole before bedrock is reached. a bent pipe is also difficult to pull.

If the driving shoe should strike the sloping side of a hard boulder, the casing might bend and cause the abandonment of a hole, though the work, if in fairly loose material, may be continued by turning the pipe slightly in the hole.

If the hole is started in top soil, the pipe, in general practice, is driven until firm ground is reached, and if in hard gravel, far enough to keep the pipe in place while drilling. Some operators, when drilling in loose gravel, drive the pipe ten or even fifteen feet before commencing to drill, but this, while permissible in some places where it has been determined that the upper gravel carries little value, would be dangerous practice to follow where the location of the values is not known, though it is often encouraged by the runner as it makes time. After the pipe has been driven to the desired depth, the core is loosened by the drill to a depth of about one foot and then pumped out. This process is repeated until all the material in the pipe, to within a few inches of the bottom, has been removed. Some operators first drill several feet before pumping, but pumping is generally done for each

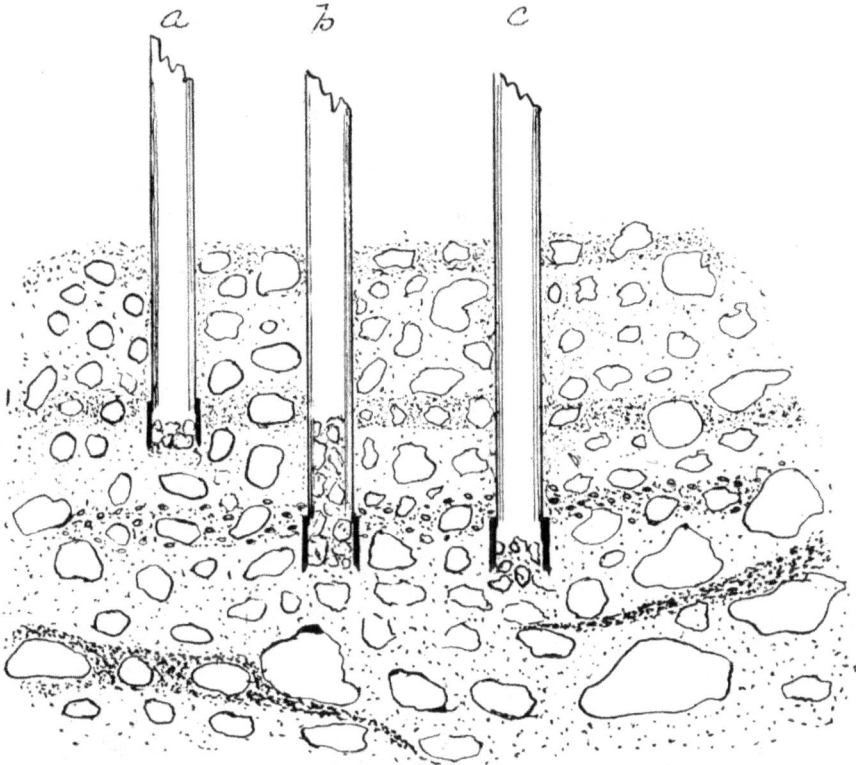

No. 17.   Drill casing.

a, after core has been drilled and drillings pumped out; *b*, core in casing after driving and before pumping; *c*, same as *a*, casing ready for next drive.

foot or less drilled, several pumpings being necessary to clean out the core.

During the drilling of a hole, a core sufficient to keep outside material from entering should be left in the pipe before and after every pumping. The depth of the core varies with the nature of the ground; it should be of sufficient length to prevent any inrush of water or material from the sides, water is also kept in the drive pipe to offset the pressure from the outside, as well as to facilitate drilling. When drilling cemented gravel or when the cutting shoe encounters hard boulders, it is sometimes necessary to drill below the casing, but in such case, after drilling, the pipe should be driven before pumping, though this is not always possible and it is occasionally necessary to pump material from below the cutting shoe; any increase in the colors at such times must be specially noted and the quantity of material should also be compared with that recovered under normal conditions.

The gold content of a placer deposit is seldom evenly distributed throughout the gravel, but is usually concentrated near the bottom, one tenth of the material drilled often containing most of the values; and the greatest care taken with the balance of the ground would be useless if an error was made while the pay streak is being drilled.

As the drill bit and stem are raised out of the hole, any adhering material is washed back into the casing, so that it may be caught with the sand pump. Two or three pumpings are generally sufficient for each foot drilled, but this depends upon the character of the material, as from some holes the pump will bring up less than one half a pan at a pumping and at others over two pans; pumping is repeated until all the drilled material is secured. When the pumpings are taken from the hole, they are emptied into a sample box, the material being caught in a pan placed in the box. The contents of the pan are then washed and any gold colors counted and recorded. The tailings from the pan are kept in a tub and rocked by the panner, as is also the material left in the sample box when a sufficient quantity has accumulated or when the panner has time.

After the core drilled has been pumped, the casing is again driven and the operations continued as before, until bedrock is reached. When bedrock is reached, repeated pumpings are necessary to remove all the drillings from the hole, and sometimes when soft bedrock is encountered the pump will loosen material that has not been drilled; the color of the drillings from bedrock is noticeable, however, and readily recognized. If the values recovered appear excessive, they may be caused by the drill hole cutting a rich seam or small pocket and the results should be noted separately from the balance of the hole.

The box for taking the pumpings is made in different sizes, sometimes 12 by 14 inches by 8 feet long. A convenient size is a box made of 2-inch planks, 20 by 20 inches by 4½ feet long. This box, held together by rods, is strong and easily transported.

The size of the rocker varies with personal ideas; it is often made 20 inches wide by 5 feet long, and is used either with or without an apron. The use of an apron sometimes facilitates rapid work, as the tailings are not rerocked. In the plain rocker, the tailings are usually rocked over one or more times, according to the amount of gold in the gravel. It is preferable to always use clean water both for panning and rocking.

The tubs in which the panning is done are either iron wash tubs or half barrels. The fine colors from the pannings and rockings are usually kept in a small dish, and at the completion of the work, amalgamated. The amalgam is put in a small vial and the quicksilver is later separated from the gold by nitric acid, after which the gold is washed, dried, and weighed. Some operators use no quicksilver and separate the black sand from the gold by close panning and a magnet, after which the gold is dried and weighed.

If the gold does not readily amalgamate, it should be noted in the log-

No. 18. Prospecting with Keystone drill, Shasta County.

book and the cause ascertained; it may be due to a coating of iron oxide, the presence of arsenic, grease, or some other cause. Sometimes so-called rusty gold will amalgamate after rubbing it with the finger against the gold pan. While it is often necessary in cold weather to use warm water to facilitate amalgamation, it is best to carry on the operation as nearly as possible under the same temperature as it would be on the dredge. It might lead to error in calculation of values if methods employed in saving the gold in sampling could not be duplicated or equaled on the dredge. The physical characteristics of the gold determine in a great measure the percentage of recovery made by

FIELD LOG

No. 19.　Time and Field Log Records.

PROSPECTING LOG-BOOK

NO. OF HOLE _10_   DATE _Nov. 12_   190

_____ CONTRACT OR OPTION

_____ TRACT OR CLAIM

FT. PROSPECTED @
EMPLOYEES

NAME	OCCUPATION
	PANNER
	DRILLMAN
	HELPER
	LABORER

DESCRIPTION

FT.

Hydraulic tailing

Coarse gravel, much sand, pay streak Hrd. granite

SIZE OF SHAFT _7½_
WEIGHT OF GOLD _13 b 4 mg_
VALUE OF GOLD _$16 an oz_
VALUE PER CU. YD. _75¢_
CUBIC FT. IN TEST _2.43_
WATER LEVEL _5_ FT.
TOTAL DEPTH _9-2_ FT.
BEDROCK _9_ FT.
LOCATION _____

REMARKS

From 4 to 7 ft. 7 in. some very fine gold.
8 ft. 9 in. to 9 ft. gold, coarse.
Some bedrock at 8'9", hard granite at 9 ft. 2 in.
From 8 ft 4 in. ground very tight, hard to drive pipe. Pipe stuck in bedrock and had to be left in hole.

TOTAL DEPTH   PROSPECTOR

No. 20. Prospecting Log Book.

the gold-saving tables of the dredge; when much light, thin, scaly, and flour gold is encountered in prospecting, it should be carefully noted, as there might be difficulty in saving same in handling large quantities of sand and gravel.

The black sand which is caught in the pan and riffles with the gold seldom contains anything of value. There is, however, a possibility of finding platinum and a variety of minerals of a heavy specific gravity, and it is always advisable, in a new district, to investigate the black sand carefully.

The practice of inexperienced engineers to use the fire assay in determining the gold content of gravel areas is more widespread than would be imagined; numerous reports submitted to dredge operators being based upon such methods which, needless to say, are worthless.

In prospecting with drills or shafts it is necessary to keep a systematic set of logs for references and calculations. These logs should give all the necessary information in as simple and comprehensive a manner as possible. Some of the largest dredging companies in California have used the Field, Time, and Prospecting log-book, as reproduced on pages 28 and 29, for a number of years, and find that they answer every purpose.

The Field and Time logs, which are kept in the field, usually by the panner or the man in charge of the drill, are made in convenient size books, about 4¼ inches wide by 7 inches long, containing 50 leaves each. The paper is made of a tough, strong quality, not easily torn when wet, and the books can be carried in the pocket and readily used in stormy weather. It facilitates matters to have them made in different colors, the Field log being usually of green, and the Time log of yellow papers. The leaves of the Field log are perforated and are torn out at the completion of each hole, and the information copied into the Prospecting log-book. In the columns giving the sizes of the colors, in the Field log, the largest are usually marked under No. 1 size, and the smallest under No. 3, but this method of grading the colors can, of course, be changed to suit any desired arrangement. The columns for depth and core indicate depth after pumpings and the length of the core left in the drill hole after pumping. These figures added give the total depth of pipe, except where an inrush of material would give an excess of core which would be noted under remarks.

In the Time log is kept a record of all work done and also difficulties and delays. It furnishes a useful record for estimating the time required for prospecting similar ground, and for promoters and investors who are anxious for speed in prospecting a record of this kind would be an education.

The Prospecting log-book is made 5 inches wide and 12 inches long, with perforated leaves, and is of sufficient length for holes up to 60 feet in depth. It should give a summary of the Field and Time log, as well as all other information pertaining to drill holes or shafts.

No average speed can be given for test drilling and as elsewhere mentioned speed is subservient to accuracy in prospecting placer ground. However, some idea can be gained from time taken during actual tests. In loose gravel as much as 20 feet was drilled in 12 hours, and at another hole on the same property in cemented gravel, a headway of only 2 feet was made in the same time. In testing shallow ground, the frequent moves and resetting require time, depending upon the distance between the holes, and the difficulty in moving.

The speed of drilling generally decreases with depth, this is caused by the time required to raise and lower the tools, the added friction in driving, and the difficulty in pulling the casing; sometimes it takes longer to pull the casing than to drill the hole, and it is occasionally found economical to leave casing that is difficult to pull.

In drilling on one property, the average depth per day, including delays caused by bad weather, moving, etc., over a total period of thirty-three 10-hour days, was approximately 12.6 feet per day; the holes averaged 24.9 feet deep, the character of the deposit was medium coarse gravel, free from clay and overlain by hydraulic tailings. The wages of the drill crew were $12.25 per day, so the labor cost in the above examination would apparently be $1.01 per foot drilled. On another examination three 70-foot holes required eight 24-hour days, including pulling pipe, moving, etc., and the cost for labor was $1.71 per foot.

The following figures are from an examination undertaken by a different engineer from the above, in rather loose deposit of sand and gravel, the boulders of granite being easily broken by the drill. Twenty-four holes of an average depth of 23¼ feet, a total of 558 feet, were sunk in twenty-six 9-hour shifts, including all delays, etc.; the cost while at work on the property being a little under one dollar a foot; however, as some of the men were brought from a distance and their time and expenses paid while traveling, the total cost of the examination, exclusive of the engineer's fee, would exceed the amount given per foot drilled.

The cost of drilling varies greatly with conditions encountered. The drill crew consists of one drillman, or drill runner; one fireman, or helper; one waterman and team; and one panner. When working more than one shift, one waterman is usually able to supply the drill with fuel and water and help in moving the drill and tools from one hole to another. The panner generally works during the day time only, his place being taken by a watchman, who keeps the Time log, during the night shift. However, this depends to a certain extent on the depth of the hole and the frequency of moving. The wages of the drill crew are usually as follows:

Drillman	$3.00 to $3.50 per day
Helper	2.50 to 3.00 per day
White panner	3.50 to 5.00 per day
Chinese panner	1.50 to 2.00 per day
Waterman and team	4.00 to 5.00 per day

Aside from the wages of the drill crew, other costs, such as fuel, repairs, maintenance and hire or purchase of drill, also figure in total costs. In the winter months heavy rains would increase the cost of transportation, making moving difficult, and greatly delay work. A serious accident to the drill machinery might also delay the examination for several days and increase the cost considerably. It is generally figured that the cost of drilling runs from $1.50 to $2.50 per foot under favorable conditions.

Too much stress can not be laid upon the care necessary in drilling operations in order to secure a sample that would represent as nearly as possible the value per cubic yard of each place sampled. If the work has been properly done in every detail, an estimate of the gold content of the property can be made with some degree of accuracy; if, on the other hand, the work has been carelessly done by unreliable or inexperienced men, the results obtained would be worthless.

It is generally conceded that the careful methods employed in washing the material sampled will equal or exceed the recovery of the gold-saving tables of the dredge, so each hole may be taken as indicative of the value of that portion of ground where it is sunk. The placing of holes in relation to the character of the deposit is an important feature, for though the mechanical details of sampling each

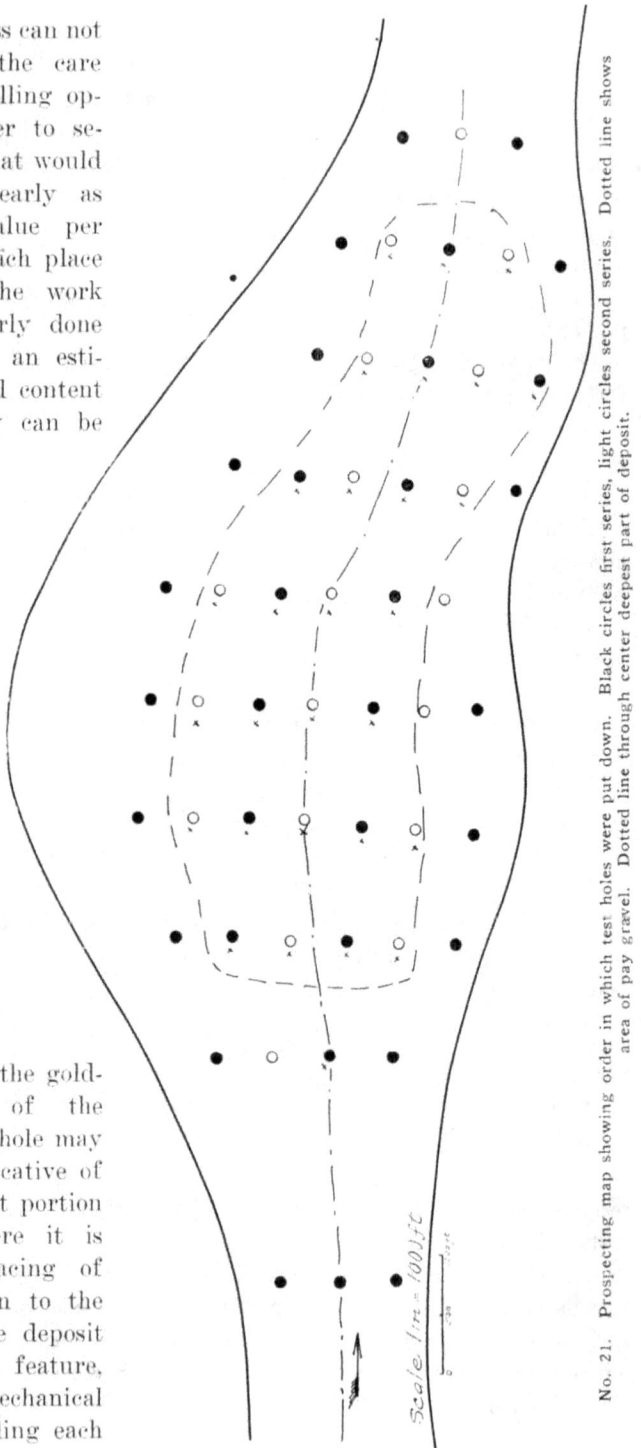

No. 21. Prospecting map showing order in which test holes were put down. Black circles first series, light circles second series. Dotted line shows area of pay gravel. Dotted line through center deepest part of deposit.

Scale 1 in. = 100 ft.

hole be most carefully done, the average values obtained might be far from representing the gold content of area under consideration.

In preliminary examinations and in prospecting large areas, a few holes are first sunk to determine the probable value and extent of the workable ground. The preliminary holes should determine if the gold occurs in wide or narrow channels or if it is fairly evenly distributed. The ground under examination is then divided into sections or squares and holes proportioned to each section.

The location and distances apart that samples should be taken can be determined only after a study of the ground. There is no rule as to the number of holes to be sunk, but on ground which shows uneven or spotted values, more tests will be necessary than where fairly uniform results are obtained, and, in general, more drill holes are required than shafts. Great economy of time is possible when the work is carefully planned in advance, and if the sampling crew consists of men well trained and directed.

The quantity of material recovered per lineal foot drilled varies greatly with different material. In drilling compact gravel, the volume of core will be greater than where a number of hard, coarse pebbles are encountered. In drilling sand or fine gravel containing much water, an excess of material may be pumped or if the pipe is driven too far ahead of drilling, it may become clogged, and too little material be secured. Some operators use a measuring box, holding one cubic foot, and measure the material as it goes to the rocker, figuring that 20 cubic feet actual recovery is equal to 1 cubic yard, theoretical quantity of core drilled. This factor was obtained from results of numerous experiments on one field and can not be applied as a general rule.

Tests in other fields have demonstrated that the actual recovery of material from drill holes, in compact gravel, has at times been 95 per cent of the theoretical quantity; and in running ground, often in excess of the theoretical core drilled. Good practice is to measure the total quantity of material pumped from each hole, and by comparing the results from all holes drilled in similar material, on an examination, a fair average can be secured and used as a check.

A simple method of making this test is to dig a hole at the end of the rocker about 3 feet square and of sufficient depth to hold all material from the hole to be drilled; all the tailing from the rocker run into this hole are allowed to settle, and the cubic contents determined and tabulated.

A dry ground containing small gravel with considerable sand would give closer results in drilling in relation to the theoretical content that should be obtained by the casing than other kinds of gravel. The theoretical quantity that should be recovered in drill tests would be the contents of a cylinder having a base the diameter of the cutting shoe and a length equal to the depth drilled.

The area of a circle the size of the cutting shoe is about .3 of a square foot, so for every lineal foot drilled in depth there would be excavated .3 of a cubic foot, or .011 cubic yards, which, as the cutting shoe soon becomes worn, is taken as .01 cubic yard, or 1 cubic yard for each 100 feet drilled. Using this factor, the method of determining values is as follows:

The gold recovered from a drill hole is reduced to value in cents; if in a new district, the fineness of gold must be first determined. The value in cents is divided by number of feet drilled and the result multiplied by 100, which gives value per cubic yard. A less simple method of arriving at the same results is by using the factor .27. Multiply the depth drilled by .27 and divide the recovery of gold in cents by the result, to get value per cubic foot; multiply by 27 to get value per cubic yard.

One operator, while agreeing with results obtained by the use of these factors when drilling compact gravel, believes that different factors are necessary when holes are sunk in other material and uses the following factors, having in numerous tests demonstrated to his own satisfaction the correctness of same:

(a) For compact gravel, the factor is _____ .01
(b) For medium gravel, the factor is _____ .011
(c) For loose gravel, the factor is _____ .012
(d) For loose gravel and sand with much water, the factor is _____ .013

To calculate with above formula multiply feet drilled by factor selected and divide value of gold from sample in cents by the result which would give value per cubic yard; as by using the different factors on the same hole, results would vary considerably, the correct selection of the factor would be a matter of much importance and could only be determined by experience.

The first two methods and factor (a) of the third method will give the same results in practice as shown by the following, taking a hole 40 feet deep that returned 15 cents gold value:

First method:

$(15 \div 40) \times 100 = .375 \times 100 = 37.5$ cents per cubic yard.

Second method:

$15 \div (40 \times .27) \times 27 = (15 \div 10.80) \times 27 = 1.389 \times 27 = 37.5$ cents per cubic yard.

Third method:

(a) $15 \div (40 \times .01) = 15 \div .40 = 37.5$ cents per cubic yard.

If factor (c) was used, however, the result would be

$15 \div (40 \times .012) = 15 \div .48 = 31.2$ cents per cubic yard.

To estimate the value of the area under examination, the value per cubic yard as found above is multiplied by the depth of the hole in feet and the added product from all holes in foot cents is divided by the sum of the depth of all holes in feet. Sometimes it is found advisable to segregate certain sections of a property under examination, and it may be possible, by cutting out a certain portion, that it is feasible to eliminate, perhaps, at one end or side of a property that the remainder would be of sufficient area and value to pay to work, whereas if all the holes drilled were averaged, the total value shown might not be attractive. There are instances in California where a portion of a property under examination was proven to be a payable dredging proposition, but the necessity of having to purchase a large unproductive acreage in order to secure the payable portion made it unattractive as a dredging investment.

In estimating the value of the gold contained in a dredging area, some engineers take a percentage, generally from 75 per cent to 80 per cent, of the total value indicated by the drill prospecting, as the amount of the gold content recoverable

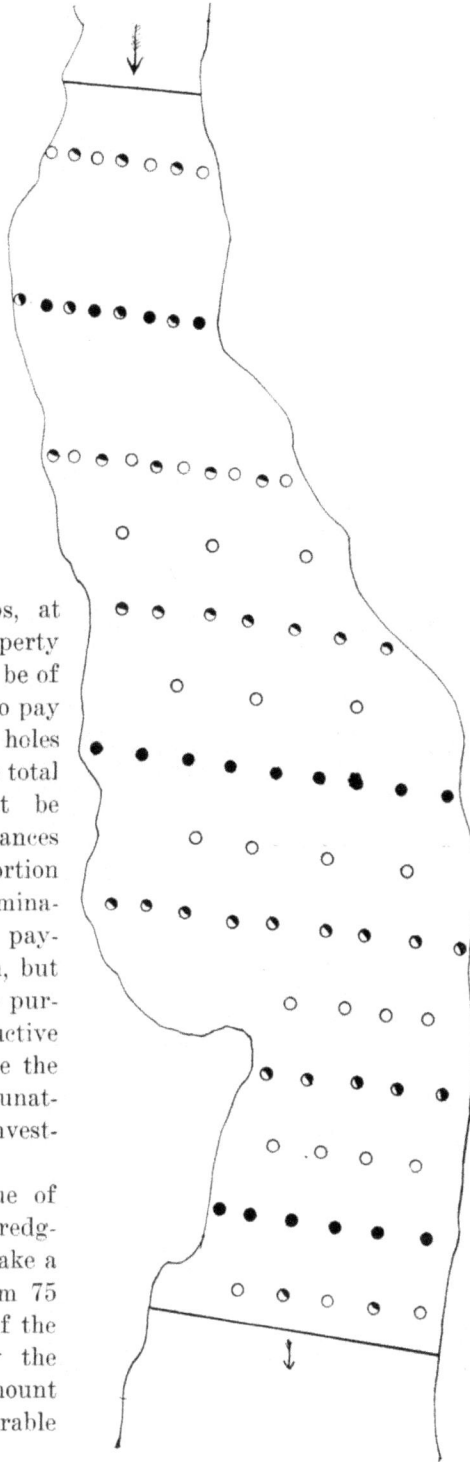

No. 22. Prospecting map showing the order in which drill holes were put down. Black circles first series, half black circles second and light circles last series. Scale one and one half inch to mile.

by dredging. These figures are based upon several tests made in the Oroville field and are also percentages of the prospecting value recovered from certain dredged areas. The consensus of opinion of the largest dredge operators in California, who have had numerous tests made, is, however, that it is impossible to give any fixed percentage to offset the various conditions of operation.

The reliability of the tests made and the accuracy of the average value indicated by either drill or shafts, depends upon the care taken in prospecting, the number of samples secured, the location of the holes in relation to the character of the deposit, and upon the experience and standing of the man in charge. Work done under these conditions should require no discounting in estimating the value of a property,

No. 23. Continental dredge, ten years and four months in operation. Just before closing down November, 1909.

other than to recognize that the recovery of the dredge is always less than the gold content indicated by prospecting and to figure on sufficient return on the capital invested to compensate for any risks taken.

After estimating the total gold contents of the property, it is then necessary to determine if the returns over working expenses and amortization of capital invested will allow sufficient dividends to make it an attractive investment.

The life of dredging propositions differs from that of vein mines in that they can be closely figured, and unlike the latter industry where the profit in sight is figured as a guarantee for the return of only a part of the capital invested, the redemption of the cost of the property and equipment must be allowed for during the life in sight, which is usually determined by having the dredge equipment of sufficient capacity to turn over the ground in ten years, as the life of a dredge with a wooden hull is generally figured at this length of time.

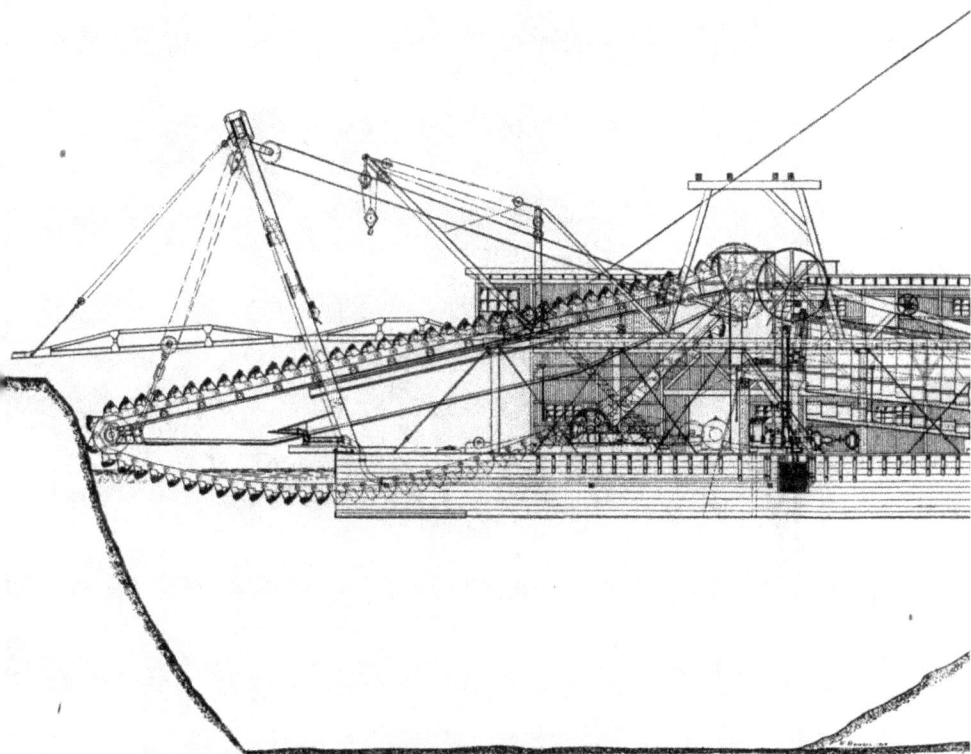

No. 24. Side elevation of Folsom Dredge No. 6, owned by Natomas Consolidated of Califo

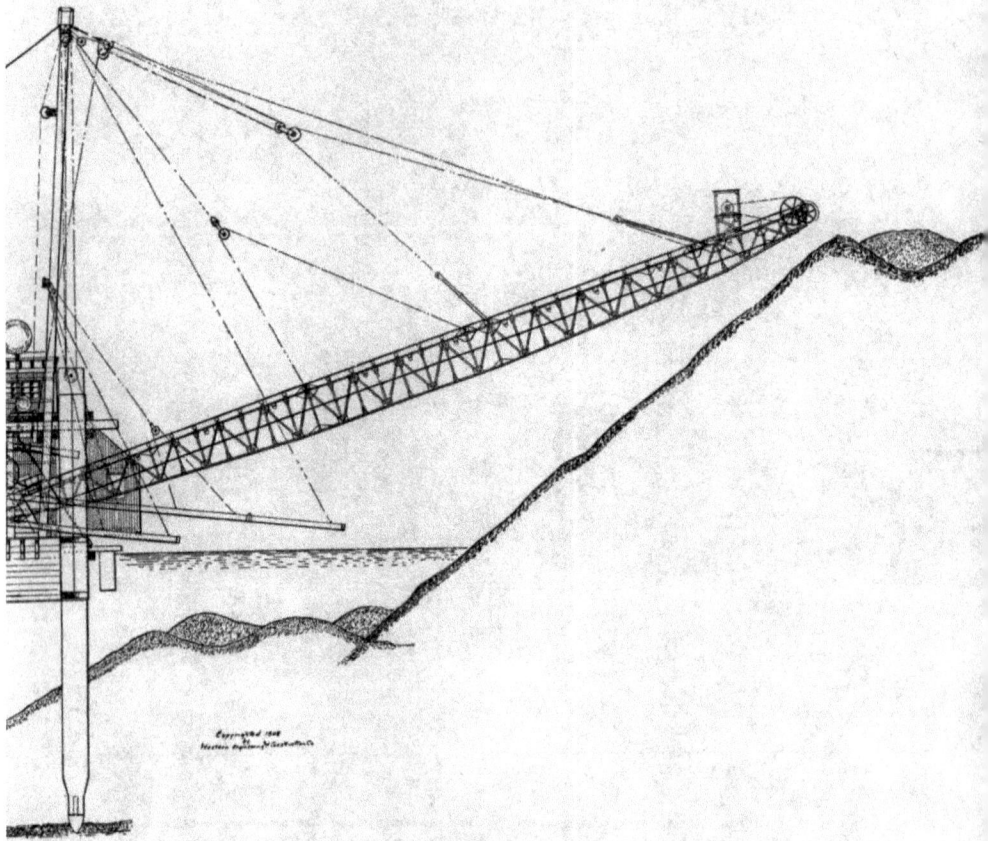

1. 9-cubic-foot buckets and double-bank gold-saving tables. See pages 190 to 195.

While the rate of interest varies with the ideas of different operators, it is generally conceded that 10 per cent of interest is the minimum to be figured in dredging propositions besides allowing for amortization of capital.

Taking the working results of different size dredges over a considerable period, the following will give some idea of the time which would be required to work a given area of ground by different size dredges and the working costs of same under average conditions in California:

An acre contains 4,840 square yards, so an acre 11 yards deep would contain 53,240 cubic yards, which to allow for places lost in turning and uneven bedrock can be figured at 50,000 cubic yards. If an examination of a dredging proposition determined the existence of 500 acres of workable ground 11 yards deep, the contents in cubic yards would be figured at 27,500,000.

A 13½-cubic-foot boat would work out about 500 acres 11 yards deep, or 27,500,000 cubic yards of medium compact gravel, in about ten years, allowing 80 per cent working time at a total operating cost of 3 cents to 4 cents per cubic yard. To work out this same area, two 7-cubic-foot boats would require nine years at a cost of from 4½ cents to 7 cents per cubic yard, and three 5-cubic-foot boats would require eleven years at a cost of from 5½ cents to 8½ cents per cubic yard. The costs given per cubic yard include total working and administration charges. These figures being taken from actual cost sheets of different dredging companies, operating the different sized dredges in California.

III.   DREDGE CONSTRUCTION AND OPERATION.

The standard placer mining bucket-elevator dredge in use in California consists, in a general way, of a wooden hull or pontoon built on the lines of a rectangle with the forward corners slightly curved.   An opening or well, through the center of the bow or forward end, extends back to the middle of the hull, where a superstructure or middle gantry supports the upper end of the ladder; the lower end being supported by cables which pass over sheaves on the front gantry to a drum or winch, so that the ladder may be raised or lowered.   A line of buckets, which excavates the gravel, is mounted on this ladder and operates through the well, rollers being fixed on the upper side of the ladder to carry the buckets.   The buckets fill with gravel as they pass around the tumbler at the lower end of the ladder, and carrying the material to a height of about 25 feet above the deck of the dredge, dump when passing over the upper tumbler, which is mounted at the top of the middle gantry. The bucket-line is revolved by the upper tumbler, through a train of gearing mounted on the sides of the upper tumbler and belted to a motor.

The gravel from the buckets is dumped into a hopper, water jets being placed so as to direct streams of water against the sides and bottoms of the buckets, thereby freeing any adhering fine gravel and gold. From the hopper the gravel and water pass to a revolving or shaking screen, which separates the coarse gravel from the fine, additional water from perforated spraying pipes extending nearly the full length of the screens, being supplied under pressure to the traveling gravel.   The coarse material passes over the lower end of the screen and is stacked behind the dredge by means of a belt conveyor, the gold-bearing material passes through the perforations of the screen into a distributor from which it flows over a series of riffle sluices or gold-saving tables on which mercury is sprinkled to amalgamate and save the gold; the fine gravel and sand from the gold-saving tables pass into side or tail sluices extending well behind the dredge.

The rapid growth of the dredging industry in California, Montana, Idaho, Colorado, Oregon, and Alaska is particularly marked by the development of the large dredge and the use of the close-connected bucket-line and digging spuds in preference to the open-connected bucket-line and headlines, as formerly used in California and on the small New Zealand dredges.

The single-lift dredges on which the upper tumbler was placed at a

No. 25.   Yuba Construction Company, Marysville, Cal.   Interior of machine shop, showing traveling crane; capacity 30,000 pounds.

sufficient height above the deck to allow the buckets to dump directly into a long sluice extending well behind the stern and supported on a pontoon, and the double-lift dredges, which dumped the material from the buckets into a sump or water-tight compartment built in the hull of the dredge from which it was raised by means of a suction pump to a straight sluice, supported by an auxiliary small scow, and extending at times more than 75 feet behind the dredge, are now considered obsolete machines in California.

In giving a general description of the California dredge it will not be necessary for the purpose of this work to consider the various mechan-

No. 26.   8-cubic-foot close-connected buckets loaded with gravel.

ical differences in the construction of machinery furnished by different manufacturers. The following description, therefore, is of the standard type of machinery in use on the majority of dredges in California. Among the principal manufacturers designing and furnishing machinery for California dredges, are the Yuba Construction Company, Bucyrus Company, Marion Steam Shovel Company, Risdon Iron Works, Link-Belt Company, Union Iron Works, and Golden State Miners' Iron Works. The principal designers and constructors of California dredges are the Western Engineering and Construction Company, Yuba Construction Company, Risdon Iron Works.

In the development of the California dredge, the digging end of the machine was first developed, the rest of the machinery increasing in size and strength in order to withstand the additional strain caused by the increased capacity of the buckets. The weight of the buckets has increased from 500 to over 4,000 pounds each, and the machinery of the dredge, as a whole, from about 150,000 pounds to over 1,500,000 pounds in the larger machines, while the capacity of the dredge has been increased from about 20,000 to over 300,000 cubic yards per month.

Dredge operators in California prefer the close-connected bucket-line to the open-connected type. In practice the close-connected line dumps at the rate of from 18 to 22 buckets per minute, while the open-connected dumps on an average of from 12 to 14 buckets per minute. Under ordinary conditions the close-connected bucket, if properly

No. 27.   Lower tumbler and bucket-line, improved Risdon type dredge.

designed, will fill to better advantage and handle boulders as well as the open-connected line, which in hard digging gravel imparts considerable vibration to the dredge on account of the intermittent spaces between the buckets. On some of the older boats in the Oroville district the bucket-line was changed to close-connected, thereby nearly doubling the yardage per month.

The following is an example of the results obtained by changing a 3½ open-connected to a 3½ close-connected bucket line—the dredge being operated eighteen months under each condition, and showing 34.8 per cent increase in yardage handled and 32.8 per cent decrease in power consumption per yard dug.

It must be always borne in mind that a dredge should be designed to suit the ground in which it is to work. Buckets suitable for free running gravel and sand, with a total absence of clay, would not be the best for either hard cemented gravel or ground containing much sticky clay. It is difficult to dump sticky material out of a long narrow bucket, and the best design where there is much clay is a wide-mouth bucket.

The buckets in use in California vary in capacity from 3 cubic feet to 13½ cubic feet and will probably soon be made 15 cubic feet; they are usually made in three pieces, consisting of bucket bottom, hood, and wearing lips. The bucket bottom is made of a special steel, and is provided with two or three eyes at the forward end and one or two at the

No. 28.   13-cubic-foot buckets Folsom No. 4 Dredge, Folsom District.

rear to receive the bucket pin. The hood is made of either cast or pressed steel. The cast steel hoods are, to some extent, used on the smaller size buckets, and the pressed steel hoods on all sizes. Whenever it is necessary to reduce the weight of the machinery, pressed hoods are used for the buckets. The wearing lips are of manganese steel, and approximately 12 inches deep.

The use of cast steel or pressed plate hoods is largely a matter of preference. The cast steel hood is designed with a shoulder for the lip to rest on and a shoulder at the lower part of the hood to rest on the back flange of the bucket bottom. This design does away with the necessity of many rivets. The use of pressed steel or cast steel hoods is also based on economic ideas depending on the character of ground the dredge is digging.

A design of bucket in which the hood and bottom are cast in one piece has been made and is being put into service, but has not been in use long enough to determine its value.

Some operators believe that it is better practice to discard the hood when the first set of lips is worn out than to replace an entirely new set of lips and hoods on the partly worn bottoms. This is more feasible with pressed than with cast hoods, as the first cost of pressed hoods is less, and good cast hoods are rarely worn out during the life of the first set of lips, though pressed hoods often are.

The most satisfactory bucket would be one with long life, all the parts of which would outwear their usefulness at about the same time, as better results are obtained with a complete new bucket-line than in running new and old buckets together. The bucket in general use is designed with two forward and one rear eye; the use of bucket bottoms with three eyes in front and two behind has been confined largely to the Yuba and Natoma dredges. For a time this type was popular with certain operators, but it is now generally admitted that the two-eye type is better and more economical. The mechanical trouble with three-eye bottoms has been in the cracking of the steel between the two back eyes. This

No. 29. Open-link bucket-line, Risdon type, 5-cubic-foot buckets.

is probably due to the middle eye of the following bucket, which fits between the two back eyes, having a tendency to pry the back eyes apart when the pins and bushings are worn. This may be caused by the swing of the long bucket-line on deep digging dredges after passing over the upper tumbler or when the buckets ride up on the lower tumbler flanges, as occasionally happens. Another objection to three-eye buckets is the increase in weight; for example, a 7½-cubic-foot bucket of this type weighs from 250 to 350 pounds more than the corresponding size in a two-eye bucket, it is also claimed that it does not wear as long, costs more to finish, and causes more wear on the upper tumbler plate than the two-eye bucket.

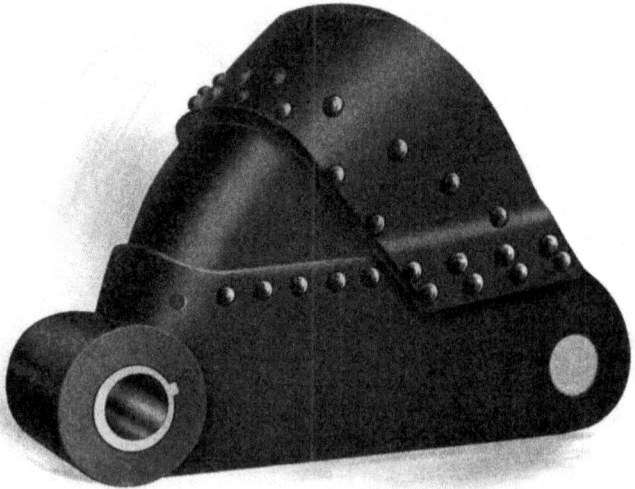

No. 30.   Two-eye buckets, showing two forward eyes and one rear eye.

The bucket pin is made of high carbon or nickel steel forged with a heavy lug or head on one end, which fits into a recess in the bucket bottom to prevent the pin from turning; a rivet placed through this lug secures the pin to the bucket bottom. In the rear eye of each bucket is a manganese steel bushing to take the wear of the pin and lengthen the life of the bucket.

The ladder carrying the bucket-line is constructed of steel angles and plates, either of the trussed or plate-girder type. The upper end is hung on the middle gantry and the lower end carries the lower tumbler bearings; the dredge has its maximum digging depth when the ladder is at an angle of 45 degrees.

Some of the ladders are equipped with a spillway, which consists of a plate fitted between the sides just below the rollers and forming a trough the full length of the ladder, thus returning the spill of the buckets to the pond where it can be again picked up. The rollers placed

No. 31.   Renewing bucket lips and hoods.

upon the ladder to support the bucket-line are usually made of high carbon cast steel and run in self-aligning bearings provided with a packing gland for excluding dirt and grit. The rollers are fastened to their shafts under heavy pressure and keyed at both ends.

The lower tumbler is six-sided, made in one piece of open-hearth steel casting, and is fitted with replaceable manard manganese wearing plates. The inside of the flanges opposite the eyes of the buckets are lined with replaceable manganese steel wearing tips. It is claimed that the one-piece tumbler type gives better satisfaction in practice than the two-piece tumbler, which is made in right and left halves bolted together. The shaft is keyed into the tumbler, the ends being supported by heavy self-aligning bearings, lined with renewable bushings and thrust washers to take up the wear caused by the side thrust. To hold the shaft in place and to exclude grit and dirt, the outer ends of the bearings are enclosed by steel caps, while on the inner side renewable packing rings exclude the grit and retain the oil for lubrication. A heavy tie-rod extends through the hollow tumbler shaft from one ladder-end casting to the other, being bolted on the outside of the cap on both sides.

No. 32. 5-cubic-foot buckets, close-connected, Bucyrus type, showing pin and bushing.

There are several types of upper tumblers in use in California, most of which are made with six sides or tread surfaces.

In one type the body of the tumbler is cast of open-hearth steel, the tread surfaces having machined grooves, parallel to the shaft, for receiving the tongue of nickel steel protecting cushion plate. These wearing plates, to which also the guide lugs or ears are cast, are fastened to the body casting by countersunk bolts or rivets.

Another make of tumbler has the guide lugs cast in one piece with the body casting, each wearing face and lug being protected by manganese tread plates bolted or riveted to the casting. A round hole is bored in

No. 33. Buckets manufactured by the Risdon Iron Works, San Francisco.

the body casting, of both types, for the upper tumbler shaft, which is pressed in and keyed to the casting.

The cushion plate tumbler is the most expensive of the two types and the nickel steel wears more quickly than the manganese wearing plates, but causes less wear on the bucket bottoms than the latter, which fact, some operators claim, compensates for the increased price and shorter life of the nickel steel wearing faces. Some upper tumblers cast in one piece with the tumbler shaft have also proven satisfactory.

The upper tumbler shaft is usually driven at both ends through a double train of gearing located on the head frame of the main gantry,

No. 34.   New type close-connected Risdon buckets.

These gear trains are driven by belt from a jack shaft located on the main deck, which also drives the ladder hoist machinery.   In some instances the tumbler drive is geared directly to the digging motor, double gears having cut teeth being most frequently used.   The tumbler shaft and the intermediate are both mounted on the large foundation castings to which, also, the ladder is often suspended.

The hopper into which the buckets discharge is placed directly under the upper tumbler, and is provided with a so-called rock bottom.   This bottom consists of a receptacle below the delivery spout constructed to retain a portion of the gravel in order that all material dumped into the buckets will fall directly upon the rock bottom, which greatly reduces wear upon the steel of the hopper.   Additional nozzles are generally

No. 35.   Structural steel ladder, assembled ready for installation, plate-girder type.   Folsom No. 6 Dredge.

No. 36. Upper tumbler driving gears and foundation castings. (Marion.)

placed in the dump hopper to play streams of water into the buckets just before and after dumping. The proper amount of water to use varies with different kinds of material. It is important that the water is not used under too great a pressure, as this would cause some of the material to be lost in the well-hole, by splashing over the sides and backs of the buckets, and if too great a quantity of water is used, more water than is desirable for the best saving of the gold will pass with the material to the screen.

In spite of the best care, some gold-bearing material spills from the buckets as they

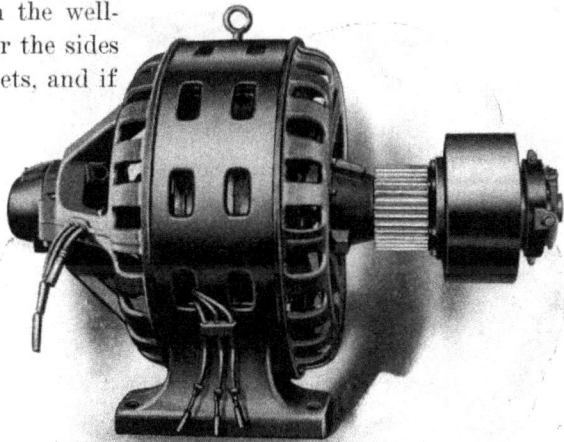

No. 37. Upper tumbler drive motor.

No. 38. Upper tumbler drive on 9-cubic-foot dredge.

No. 40. Interior of revolving screen.

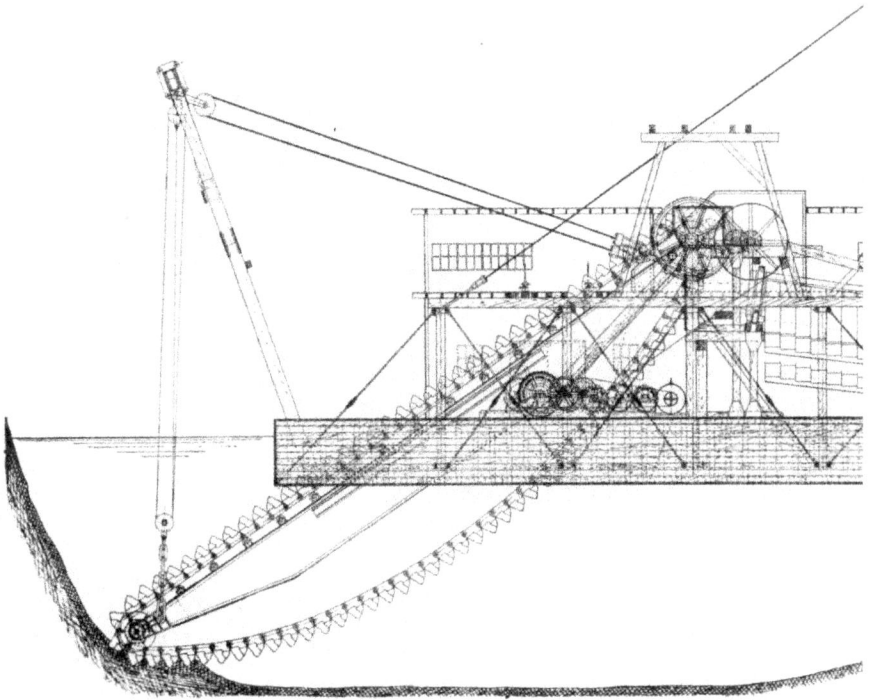

No. 39. Side elevation of electric driven placer mining elevator

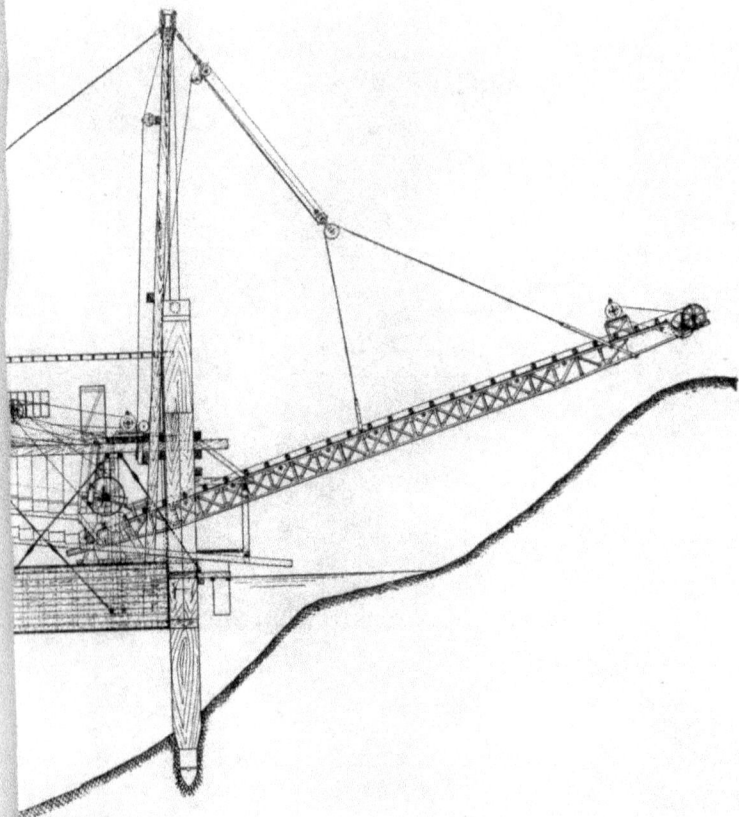

dredge. Bucyrus Company design. California type.

dump, just missing the lip of the hopper. All modern dredges are generally equipped with a save-all, consisting of a small sluice which projects into the well-hole and extends under the gold-saving tables to the stern of the boat. Steel grizzly bars, placed below the dump hopper and extending into the well-hole some distance, prevent gravel larger than $1\frac{1}{2}$-inch from going into the sluice, and dump the coarse gravel into the well-hole. As much as $700 a clean-up has been recovered from the save-all on one dredge.

The screening arrangement of the dredge was not developed as rapidly as the digging end. For a long time the dredges were digging more gravel than they could wash and screen; this defect, however, was overcome by the increased screening area and by adding more water under high pressure. The screening, as well as the digging end of the dredge, must be designed to suit the conditions under which the operations are carried on. In most operations a revolving screen is better than flat shaking-screens.

There are two types of revolving screens, the straight and stepped, the straight design being the most used. In the revolving screen the perforated shell is composed of sectional plates, so designed that the sections can be replaced when worn. Narrow steel bars are riveted to the surface of the sections to carry the gravel well upward in the rotating movement, and provision is made for the shifting of these bars so that the whole surface of the shell sections may be uniformly worn. The screen frame or skeleton is made of six longitudinal members, composed of angles bolted together in pairs, and riveted to the tubular end sections carrying the friction tires. The revolving screens vary in size according to the material to be handled, the diameter being usually about 6 feet and the length from 20 to 35 feet. The screen is carried on two cast steel friction tires, riveted to the shell plates, and in operation is supported by four heavy friction wheels or rollers, mounted on longitudinal shafts, two on each side of the screen; the two rollers at the lower end acting as drivers turn the screen through frictional contact with the tires.

In the stepped design, shoulders are formed by the successive reductions in diameters of the shelled sections, and retard the flow of the material, which, is claimed, gives a more perfect disintegration of the coarse material. These shoulders are protected by renewable manganese rings and the sections are renewable, the same as in the straight type. This type of screen was designed by Charles W. Gardner, who holds a patent on same.

The shaking screens, which are not used much on the new dredges, consist of perforated steel plates mounted on a flat frame and driven by crank shaft or eccentrics. This type of screen performs the same func-

No. 41.　Revolving screen and drive.

tion as the revolving screen, but is not so well adapted for the handling of large quantities of clay or cemented gravel, and does not wash the gravel as thoroughly as the revolving screen when clay predominates.

The diameter of the holes in the screens ranges usually from 5-16 to ⅜-inch at the upper end,

No. 42.   Revolving screen, straight type.

and ⅜ to ⅝-inch at the lower end of the screen, being governed by the character of the ground to be handled.

The screen casing and distributor, which catches the screened material and feeds it to the gold-saving tables, is a V-shaped

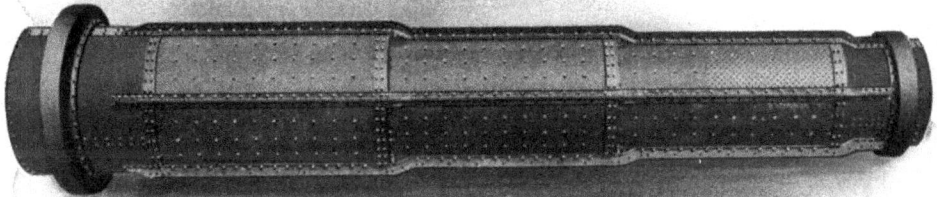

No. 43.   Revolving screen, stepped type.   Patented by Chas. W. Gardner.   Made by Marion Steam Shovel Company.

trough, in the side of which gates are fitted, one at the head of each table, for the purpose of controlling the flow of material to each separate set of tables.

The trough is lined with renewable wearing plates, water being furnished by perforated pipes inside the trough, and also outside, to assist the flow of material as it leaves the trough gates.

No. 44. Standard type of shaking screen, Indiana No. 3 Dredge.

There are two types of tailing stackers in use, the belt conveyor, with which the standard makes of dredges, except the Risdon, are equipped; and the bucket-elevator, which the Risdon Iron Works recommend with their dredge, though in their catalogue they state that they will also furnish the belt conveyor if desired. The size of the conveyor belt varies from 28 inches to 44 inches in width. The stacker is from 80 to 145 feet long and operates at an inclination of from 18 to 20 degrees.

The rubber conveyor belt requires little attention and lasts about nine months; this varies, however, considerably. On a 7-cubic-foot dredge, the service given by conveyor belts varied from 2,607 hours running time to 3,442 hours, while the dredge handled 232,000 cubic yards in the first instance, and 664,000 cubic yards in the second. On a 3-cubic-foot

No. 45. Lattice girder type tailing stacker ladder for supporting belt conveyor.

dredge, one belt lasted 3,213 hours, while the next belt used gave a service of 5,806 hours, the dredge handling over double the yardage that it did during the life of the first belt. On a 5-cubic-foot dredge, one

No. 46. 38-inch Robins tailing conveyor.

belt gave a service of 6,380 hours, the dredge washing 782,000 cubic yards of gravel, while the belt replacing this one, lasted only 3,871 hours, during which time the dredge handled 475,000 cubic yards of gravel. The character of the ground handled and local conditions gov-

No. 47. Showing large boulders which passed over tailing stacker on the Calaveras dredge,
Jenny Lind, Calaveras County, Cal.

No. 48. Bucket conveyor for stacking dredge tailing. Risdon make.

ern, to a certain extent, the life of the conveyor belt, water containing a great deal of sand and grit being decidedly unfavorable.

In the manufacture of rubber belting it seems difficult to turn out a uniform product. It has often been the case that two belts, furnished by the same manufacturer and apparently duplicates, have given an entirely different service on the same dredge. Canvas and leather belts for conveyors have proven unsatisfactory; a so-called weather-proof leather belt, guaranteed for ten months' service, was tried in the Oroville

No. 49. Ladder hoist with automatic brake.

district, and lasted only six weeks; canvas belts used lasted from three to six weeks.

The conveyor belt is easily renewed, and the loss of time for repairs on this style of stacker is less than with the bucket elevator, where there are many wearing parts, such as the buckets, bucket connecting pins, tumblers, etc., requiring repairs and renewals.

The bucket-elevator stacker has not given satisfaction on the large close-connected bucket dredges, as the weight would be excessive to get the required capacity with a long stacker. The modern dredge builders and operators, excepting as stated, used the belt conveyor. The power to drive the stacker belt is furnished by a motor placed either at the outer or lower end of the stacker frame.

There are several types of troughing idlers used with the belt conveyor. Some designs have sets of three or four idler pulleys of equal length set in such a position to concave the belt, and another type has an idler pulley in the center with two short ones set obliquely at each end, in order to raise the sides of the belt sufficiently to keep the material thereon. The latter style of troughing idlers has proven very satisfactory. The belt, returning underneath the stacker, is carried on straight rollers several inches longer than the width of the belt.

The ladder hoists are sometimes driven by a separate motor and are

No. 50.   Port winch of 9-cubic-foot dredge.   Folsom No. 6.

equipped with an automatic mechanical brake, so that the winchman, by operating only the motor controller, can raise or lower the ladder or suspend it at any angle. The drum, gears, automatic brake parts and shafting are of steel, mounted on a structural steel frame. Most ladder hoists are driven from the digging motor, the drum being fitted with two large hand-operated belt brakes, and from another pulley on the same shaft a belt is carried up overhead to the main drive. Friction clutches are employed, so that the ladder can be hoisted while digging.

The winch machinery for operating the side-lines on a modern dredge consists of eight drums, mounted in pairs on separate shafts and fitted with expanding band frictions and outside check belts. Two of these drums operate the bow-side swing lines, two, the lines for raising and lowering the spuds, and two, the stern breasting lines. The other two drums are auxiliary and can be used one for a headline and one for hoisting the stacker. At the end of each drum shaft is a drive gear, the one for the bow-swing lines being fitted with a jaw clutch, so that the other drums can be cut out and remain idle when the swing lines are in use. The gears and drums of this machinery are all made of steel and the shaft of forged steel.

No. 51. Installing a 75-foot steel spud, Folsom No. 6 Dredge.

The majority of dredges in California are operating on spuds. The extensive operations with spud and headline dredges in the Folsom and Oroville districts are said to have demonstrated clearly that the spud boat will handle more yardage and clean the bedrock better than the dredges working on the headline. In some instances the capacity of the headline boats was increased by equipping them with digging spuds. Mr. Krug, who operates five spud dredges in the Oroville district, has at different times worked these boats on headlines, and demonstrated to his own satisfaction that the yardage per month was less and that the wear and tear on the boat was very much greater with the latter method. Some operators prefer the headline in light, soft, shallow ground having a level surface, but for deep ground having an uneven surface or in compact gravel prefer the spuds.

The spud dredge is equipped with two spuds, one built of structural steel and the other of wood, sometimes both being of steel. The steel spud is used while digging, and the wood spud to step the dredge

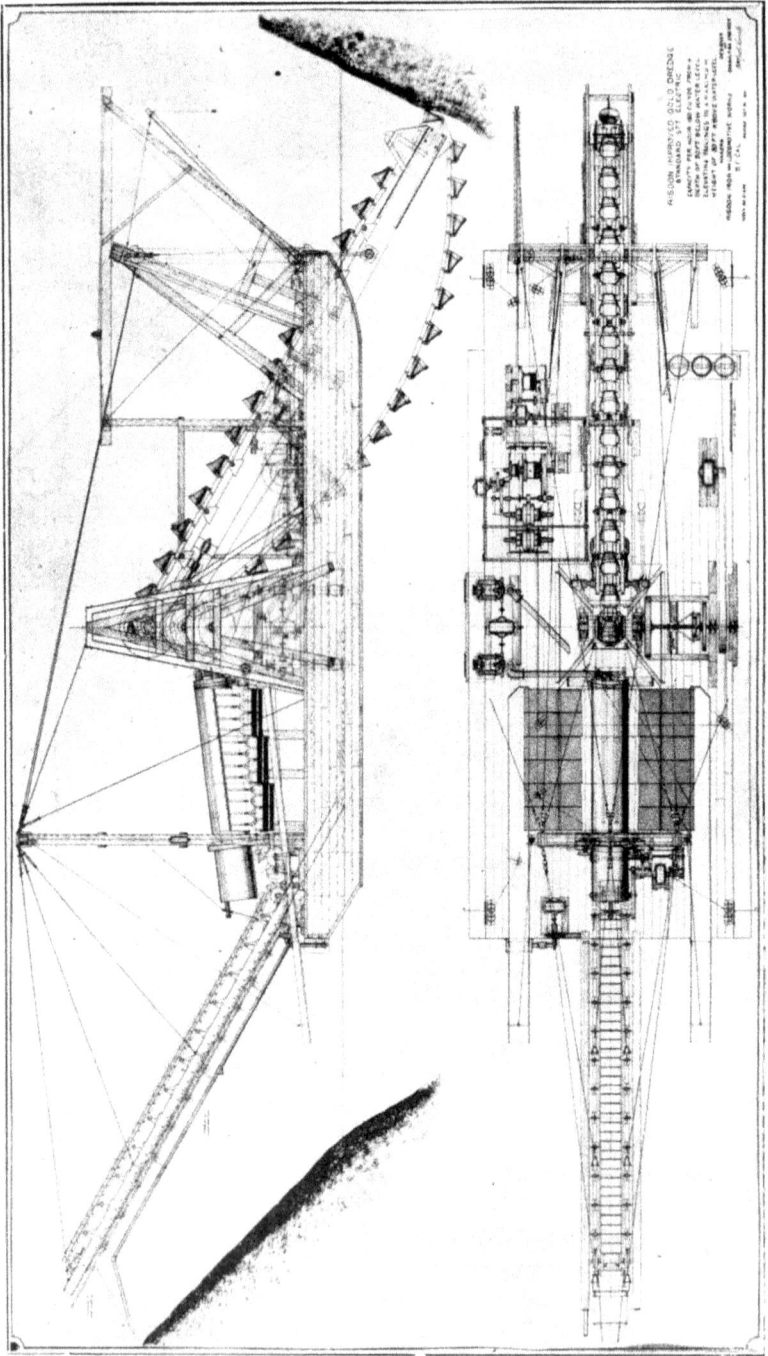

No. 52. Electrically driven Risdon open-link-bucket elevator dredge, revolving screen, bucket tailing conveyor, etc., operating on headline.

ahead when the digging spud has been raised at the completion of the cut.

The dredge is moved to right or left through an arc of a circle by means of steel wire ropes located at the bow, one fastened to the port shore and one to the starboard shore, also two steel wire ropes are placed at the stern and fastened in a similar manner to the shores.

Bucket-line in operation.   California type, spud dredge.

The headline dredge is equipped with four side-lines, similar to those on the spud dredge, and the dredge is kept against the bank, while digging, by means of a heavy steel wire rope or headline, which is fastened to the ground ahead of the dredge.

The motors and pumps vary in size, according to the capacity of the dredge. The table on page 62 shows the electric motor equipment and the size of the pumps on California elevator dredges, ranging in size of buckets from 3 cubic feet to 13½ cubic feet.

ELECTRIC MOTOR EQUIPMENT ON BUCKET ELEVATOR DREDGES IN CALIFORNIA.

NAME OF DREDGE.	Type of Dredge.*	Size of Buckets	Average Depth of Ground	Average Power of Consumption		Total Rated H. P.	Bucket Drive Motor.	Winch Motor	Washing Screen Motor.		Tailing Stacker Motor.	Stacker and Screen Motor.	Pump Motors. Gold Tables and Screens		Pump Motors. Sprays to Hopper.		Pump Motors. Sand Pump		Pump Motors. Other Pump.		Other Motors.
		Cu Ft	Ft.	H. P.	Kw.	H. P.	H. P.	H. P.	Type.	H. P.	H. P.	H. P.	Size.	H. P.	Size.	H. P.	Size	H. P.	Size.	H. P.	H. P
Indiana No. 1	Close C.	3	35			180	50	20	Shak.	20	20			40			6 in.	30			
Butte	Close C.	3½	34	101	64 5	165	50	20	Shak.			20	8 in.	40	2 in.	5	6 in	30			
Nevada	Close C	4	27			168	50	15	Rev.	15	15			40		3		30			
El Oro No 1	Close C.	5	35	137	102	290	75	30	Shak.	20	20		10 in.	100	2½ in	5	6 in.	40			
Pacific No. 3	Close C.	5	34			208	75	20	Shak.	15	15			50		3		30			
Feather No 1 (Cherokee)	Close C.	5	35	203	151.8	205	100	25	Shak.			20	8 in.	50	1½ in.	10					
Ophir	Close C	5	27			262½	100	20	Shak.	20	15		8 in	50	5 in.	7½		50			
Leggett No 3	Open C.	5	30	95		175	75	25	Rev.	10	10			50		5					
Calaveras No. 1	Close C.	5	28			208	75	20	Shak.			30	Two 7 in.	50	2 in.	3		30			
California No 2	Close C.	5	38½			195	75	20	Shak.			30	Two 6 in.	40			6 in.	30			
Exploration No. 2 (Biggs No. 2)	Close C.	5	30			238	100	20	Shak.	20	15		Two 7 in.	50	1½ in.	3	6 in.	30			
Hunter	Close C.	5	35	125		225	100	20	Rev.	20	20		8 in.	50	4 in.	15					
Lava Bed No. 2	Close C.	5	32	125		228	75	20	Shak.	20		30		50		3		30			
Empire	Close C	5	35	130		220	100	20	Rev.	15	15		8 in.	50	4 in.	15					
Gardella No. 1 (Wym. Rav.)	Open C.	5¾	10	84	63	165	50	15	Rev.	10	10			50		5					† L. H. 25
Pennsylvania	Close C.	6	28	140		220	75	20	Shak.		10			75					2 extra	40	
Yuba No. 1	Close C	6	60	258	198	237½	100	30	Rev.	30	20		8 in	50	3 in.	7.5					
Yuba No 2	Close C.	6	60	209.6	155.4	230	100	30	Shak.	20	20		8 in.	50	3 in.	10					
Yuba No. 11	Close C.	7	70	315	262	405	200	25	Rev.	35	25		10 in.	75	4 in.	10			10 in.	35	
Baggett No. 1	Open C.	7	38	110		215	75	20	Rev.	10	20			50		5					L. H. 35
Pacific No. 1	Close C	7½	34			300	100	20	Shak.	20	20		8 and 10 in.	25 and 50	4 in	15		50			
Pacific No. 4	Close C.	7½	29			300	100	20	Shak.	20	20		8 and 10 in.	75	4 in.	15	8 in.	50	Low		
Scott River No. 1	Close C.	7½	30			345	125	30	Rev.	30	30		10 in.	85					Pressure		L. H. 45
Feather No. 2	Close C	7½	27½	234	175	300	150	25	Rev.	25	15		10 in	50	4 in	10			10 in.	25	
Feather No. 3	Close C.	7½	28½	232.8	173.6	320	150	25	Rev.	35	25		10 in.	50	3 in.	10			10 in.	25	
Yuba No. 4	Close C.	7½	65	243.6	181.4	300	150	25	Rev.	25	25		8 in.	50	3 in.	10			8 in.	15	
Yuba No. 6	Close C.	7½	70	311	233	300	150	25	Rev.	25	25		8 in	50	3 in.	10			8 in.	15	
Yuba No. 10	Close C	7½	70	393	293	405	200	25	Rev.	35	25		10 in.	75	4 in	10			10 in.	35	
Natoma No. 2	Close C	8	25	322	240	430	200	25	Rev.	50	35		10 in.	75	3 in.	10			10 in.	35	
Natoma No 3	Close C.	8	42½	370	277	420	200	25	Rev.	50	25		10 in.	75	3 in.	10			10 in.	35	
Folsom No. 3	Close C.	8½	30	401.5	299.4	490	150	30	Rev.			30	14 in	100	5 in	30	Monitor	100	7 and 8 in.	50	
Folsom No. 5	Close C.	9	60	484.8	363 8	540	200	30	Rev.	30	30		14 in.	100	2 in.	30	Monitor	100	4 in.	20	
Folsom No. 6	Close C	9	60	572	428	790	200	30	Rev.	30	50		12 in.	100	5 in	30	Monitor	300	12 in.	50	
Butte Creek Cons.	Open C.	11	27			375	150	85	Rev.	15			10 in.	100						25	
Folsom No. 4	Close C.	13	20	260	195	415	200	30	Shak.	30	40		16 in	100	5 in.	15					
Natoma No. 1	Close C.	13½	19	445	333	605	300	35	Shak.	35	35		14 in.	150	5 in.	15			10 in.	35	

*Close connected and open connected bucket lines. †Ladder hoist.

No. 53. Yuba Construction Company, Marysville, Cal. Interior of machine shop.

The electrical equipment on dredges requires good care, both from a mechanical and electrical standpoint. The electrical equipment on many of the dredges in operation caused considerable trouble and expense until the temporary arrangements were abandoned and good permanent work substituted.

It is important, from an economic standpoint, that the winchman be thoroughly instructed in the handling of the controlling apparatus and operation of the motor equipment, operated from the pilot house,

No. 54. Type of switchboard used on many Oroville dredges for low voltage, motors operating at 440 volt.

as the power consumption and maintenance can through careful operation be materially reduced.

All dredges should be provided, for the winchman's information, with one ammeter on the bucket-line controller circuit, so that he can prevent overload of the motors, thus reducing the cost of power and maintenance of both the mechanical and electrical equipment. To prevent burn-outs of compensators and autotransformers, and of the constant speed motors, now used for stacker, screen, and pumps, the attendant in charge should be instructed not to throw the starter into running position before the motor reaches full speed, which is the cause of many burn-outs on the dredge. It is more satisfactory in these and other places, where a slow starting weight of heavy load is required, to use variable speed motors; for instance, where the stacker, on an inclination of from 15 to 20 degrees, is being started, loaded with tailing, the starting torque is very strong, causing a heavy current to go through motor and starter, often causing fuses, compensators, coils, and autotransformers to burn out, which will not occur if variable speed motors are used.

All motors and starting compensators should be protected from water and dirt with water-tight covers. The transmission distribution points on shore should be provided with a nonautomatic oil switch, to disconnect the three-conductor cable from the transmission line, or in places where the disconnection can not be operated from the substation, a triplex cable terminal should be installed to secure a good mechanical cable connection.

The transformer shelter on board the dredges should be well ventilated, for this purpose an exhaust should be so placed and connected to the transformer compartment as to give the air free circulation, thereby reducing the temperature of the transformers. This exhaust could also be used to ventilate the hull, which often needs better ventilation.

In order to reach the transformers in case of fire, a sliding door, opening through into the dredge from the transformer shelter, should be provided, and a box of sand placed within easy reach to be used as a fire extinguisher; water should never be used. A good automatic safety guard would be to place a box containing sand directly over the transformers, having a removable bottom of the swinging door pattern, sustained in position by a cam or level held in place by means of a light wooden strut which would burn and release the sand, thereby smothering the flames in the transformers before any serious damage could be done to the dredge.

It is important to provide the dredge with a searchlight, so placed on the front gantry, and operated from the pilot house, that the winchman can find the location for his work at night, and for shore work, such as shifting side-lines, moving power cables and sheaves on the gantry, etc. Many operators are using lanterns, which in stormy and rainy weather are difficult to handle and cause delay. For headlights, some dredges are provided with "clusters" of from five to six incandescent lamps; these, however, are expensive in maintenance and not economical in proportion to the lighting effect. On the front gantry there are from 35 to 40 incandescent lights, using from 18 to 20 amperes, which give insufficient light, are expensive to maintain, and could be replaced by a six to ten-ampere arc or searchlight, which would furnish all light needed.

Two disastrous fires have demonstrated that great care should be exercised in the operating of electrically driven dredges, as both fires were

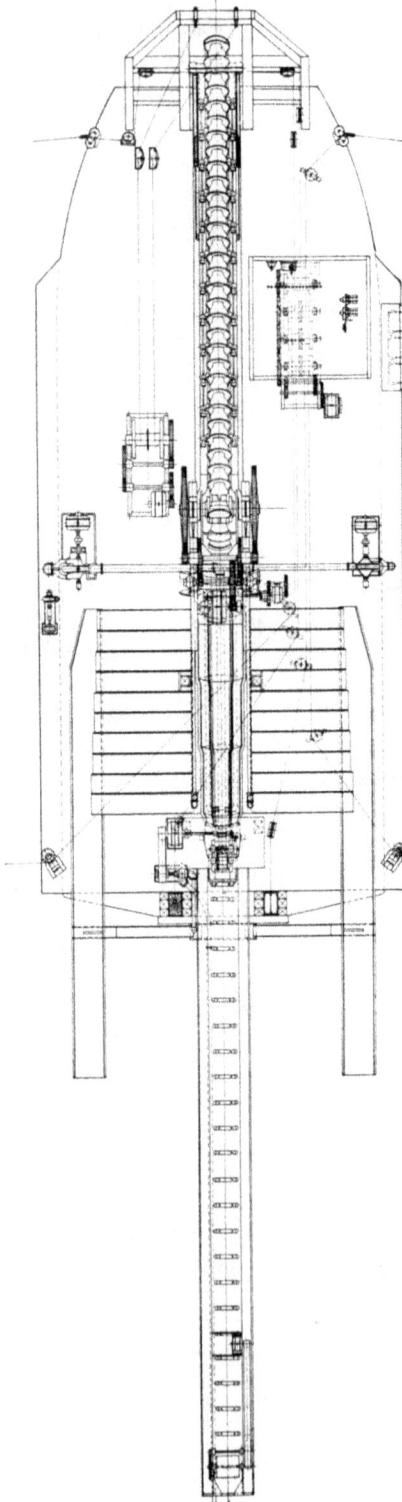

No. 55. Plan view of electric driven placer mining elevator dredge. Marion Steam Shovel Company design. California type.

caused by the electric currents. The fire on the Shasta Dredging Company's boat on Clear Creek, near Redding, started in the transformers; and on the Viloro No. 1, at Oroville, at the oil switch. This latter dredge was undergoing extensive repairs and the fire may have been caused by the listing of the hull at night, due to insufficient caulking, thereby causing a disarrangement and crossing of wires.

It is still the custom to equip dredges with sea valves so that the hull may be flooded, but in the above cases they would have been of little value, as the fire spread rapidly and the housing was quickly destroyed. Water tanks were placed on the roofs of some of the early dredges, and this practice is still being followed by many operators. The loss from fire, however, has not been great, considering the number of dredges operating.

In the Oroville and Yuba districts, the electric power is usually brought over the primary line from substation to dredge site at 4,000 volts, whence it is carried aboard dredge through shore cable and then stepped down through oil-cooled transformers to 400 or 440 volts, for all motors. In a few cases the current is used at 2,000 or 2,200 volts for all motors of 50-horsepower and over, and the transformers aboard the dredge step down from 2,000 to 440 volts for all motors under 50-horsepower.

In the Folsom district, the current is taken aboard the dredges at 2,000 or 2,200 volts, for all motors 50-horsepower and over, the smaller motors operating at 440 or 400 volts. The current is stepped down aboard the dredge by oil-cooled transformers, the primary voltage being 2,000, and the secondary 440 volts.

Some operators put the transformers on shore in place of on the dredge. Many of the dredges to be built in the near future will not require any transformers, as the leading manufacturers of electrical motor equipments are now making motors as low as 15-horsepower to operate at 2,000 volts. This will make it possible to equip a dredge with all motors designed to operate at 2,000 volts. Most power compa-

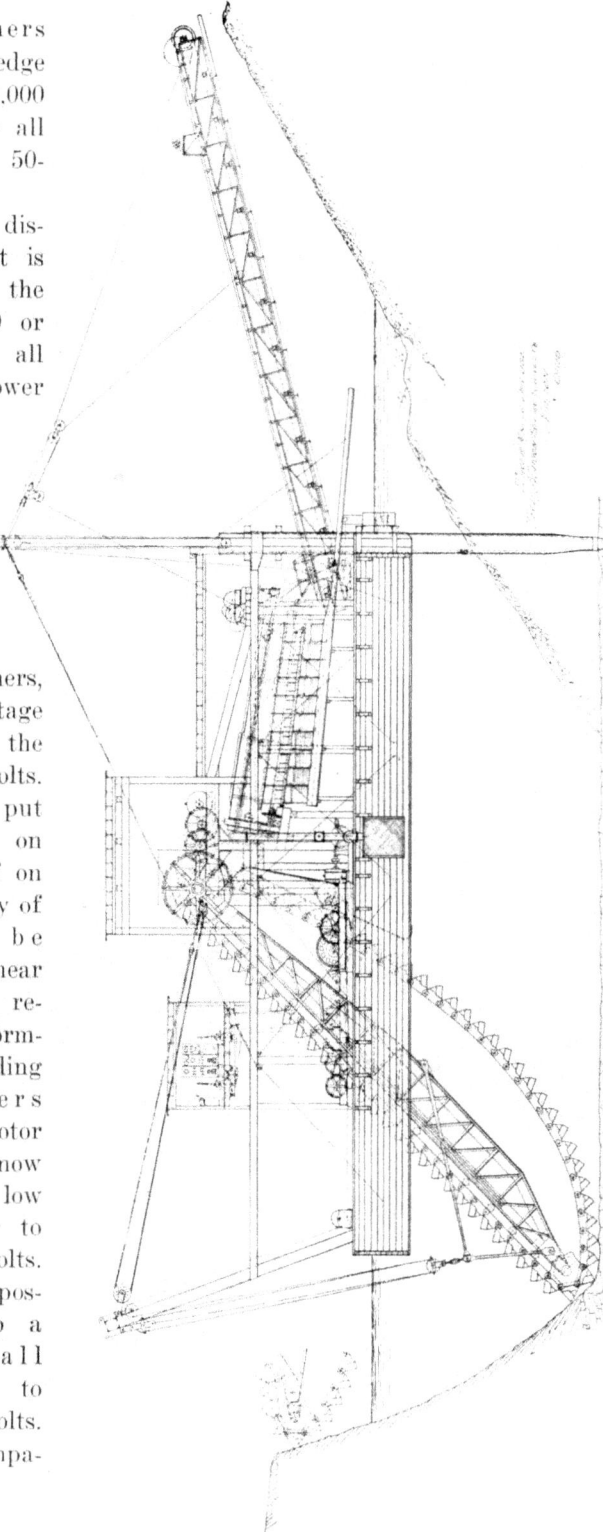

No. 56. Side elevation of electric driven placer mining elevator dredge. Marion Steam Shovel Company design. California type.

nies in California are now able to furnish electric current at 2,000 volts to the dredges through substations conveniently located along the high tension circuits. The elimination of transformers aboard the dredge is a decided advantage, inasmuch as the risk of fire is consider-

No. 57.   Viloro No. 1 Dredge, destroyed by fire September 2, 1909.   Oroville District.

ably reduced, the insurance rate lessened, and considerable weight taken off the hull. It is also a fact that all motors running at a uniform voltage of 2,000 volts afford more economical operation through the saving of transformer losses and the acquisition of better regulation.

No. 58.   Electric cable supported on barrel pontoons well arranged.

The electric power is brought aboard the dredge by means of insulated cables, usually of the submarine armored type. Some operators use floats or small pontoons to carry the cable across the pond. This is considered good practice when the floats are arranged to keep the cable out of water, but where the cable is wet and dry alternately, unless thoroughly protected by rubber hose, the insulation quickly rots in hot weather. Mr. Carr, formerly of the Oroville Dredging, Limited, says that he has found it most practicable to carry the cable under water and uses no other method, as when the cable is completely immersed the life is greatly lengthened.

All the dredge hulls in California are constructed of wood, generally Douglas fir; the size and strength of the hull depending upon the weight of machinery carried and class of gravel to be handled. In constructing the hull, a pit is usually excavated 150 feet square by 8 feet deep, the hull being built upon a temporary wooden frame about four feet high to allow room for caulking and spiking the bottom planking.

No. 59.   Electric cable supported on barrel pontoons too far apart.

When the hull is completed, the pit is filled with water and the hull floats off the construction foundation, which also floats, and is easily removed from the pond and so does not interfere with the operating of the dredge. Some dredge hulls are built at the side of the pit or river and launched into the water.

The forward section of the hull is provided with a center opening or well-hole, extending from the bow to the tumbler or middle gantry, and of sufficient width and length to allow the digging ladder and bucket-line to be lowered to an angle of 45 degrees, when digging at its maximum depth.

The size of the timber used is generally as follows: Outside and bottom planks, 4 by 12; well-hole, stern, and bow blanks, 6 by 12; deck planks, 3 by 6 or 4 by 6. The framework varies from 6 by 8 to 8 by 12 timbers, according to the size of the hull. The well-hole planking is

spiked to the framework of the hull with 12-inch spikes, and also drift
bolted together through the edge of the planks with ¾-inch by 30-inch

No. 60. Interior of dredge hull as designed and constructed by the Western Engineering
and Construction Company.

long drift bolts; the bulkheads are built in the same manner on each
side of the well midway between the well planking and the outside
framework, making four bulkheads the entire length of the hull.

No. 61. Showing bow gantry construction. Natomas No. 3 Dredge, Folsom District.

The dredge hulls built during the last five years have greatly increased in weight, and are strongly braced with two overhead trusses extending the entire length of the boat on either side of the well-hole; also an overhead truss across the hull in the center, which is attached to and distributes the loads upon the tumbler gantry, thereby stiffening the hull fore and aft and athwartships. These trusses generally consist of 14 by 14-inch posts, having 14 by 16-inch cap stringers, and are braced by heavy steel diagonal truss rods between the posts.

To prevent the forward pontoon sections on either side of the well-hole from warping and sagging, which was a fault of the earlier boats, the bow gantries have been stiffened and redesigned to form a truss. The gantry now consists of four posts 14 by 20 inches, rising about 36 feet above the deck, two being located on either side of the well-hole and two on the outside of the bow, well braced with steel rods and timber struts.

No. 62. Bow gantry construction. Hunter Dredge. Oroville District. California type.

The middle or tumbler gantry posts are 16 by 20 inches and of sufficient length to support the upper tumbler at a height of 23 to 25 feet above the deck. The stern gantry posts are 14 by 16 inches and rise about 50 feet above deck, resting on heavy timbers inside the hull.

The gantry caps for the 5-cubic-foot and 7-cubic-foot machines are usually of timber, with steel side-plates the full length of the caps and extending down the gantry posts, giving a substantial fastening to same. The larger machines are provided with structural steel caps.

The spud casings are usually built of timber 14 inches thick and of sufficient depth to take the spuds. It is considered good practice to place on the stern of the dredge a heavy steel casting for the purpose of taking up the wear and tear caused by the rubbing of the steel spud against the wooden hull. See page 189.

Steel hulls for gold dredges are receiving more attention, largely on account of the increased cost of suitable dredge lumber. Many of the dredges in South America are equipped with steel hulls. The use of steel in place of wooden hulls is due mainly to the insect life found in these parts, which is very destructive to wooden structures, and also on account of the climate, which causes the rapid decay of wood. Many prominent California dredging engineers advocate the use of steel hulls in California, claiming the climatic conditions destroy the lumber hulls too quickly; but this is probably due to defective ventilation, as it has been found necessary to place adjustable ventilators so that a current of air can be forced through the wooden hulls in order to prevent dry rot. Two California type dredges equipped with steel hulls and $8\frac{1}{2}$-cubic-foot close-connected buckets are at present under construction in California, one for Burma, India, and the other for Colombia, South America. See illustration No. 63.

The single bank of riffle sluice tables is used upon 3 to $7\frac{1}{2}$-cubic-foot dredges, having a capacity ranging from 60,000 to 125,000 cubic yards per month. On the $8\frac{1}{2}$ and $13\frac{1}{2}$-cubic-foot dredges the riffle area obtainable with the above table was found insufficient, and to meet the requirements of these larger dredges a second set of tables was placed under the single bank tables, thereby doubling the gold-saving area.

The introduction of the double-bank tables, which was developed by the Folsom Development Company in 1908, has increased the gold-saving area; and the use of four tail sluices of different lengths, extending over the stern of the dredge, thereby distributing the sand and small gravel tailing over a greater area, has made it possible to discard the use of the sand pump, which was always a source of annoyance and expense.

The construction of the single-bank tables is usually of wood, the tail sluices on this type of table extend about 20 feet over the stern of the dredge. In the construction of the double-bank tables, both the upper and lower set are made of steel. The tailing sluices from the lower bank tables extend only a short distance over the stern of the dredge, about 20 feet, while in the case of the upper bank of tables, the upper tail sluices extend a distance of nearly 40 feet over the stern of the dredge. The tables are covered with so-called Hungarian riffles, consisting either of strips of beveled soft pine, shod with $\frac{1}{8}$ by $1\frac{1}{4}$-inch strapiron or of angle-irons 1 by 1 by 3-16-inch riveted together.

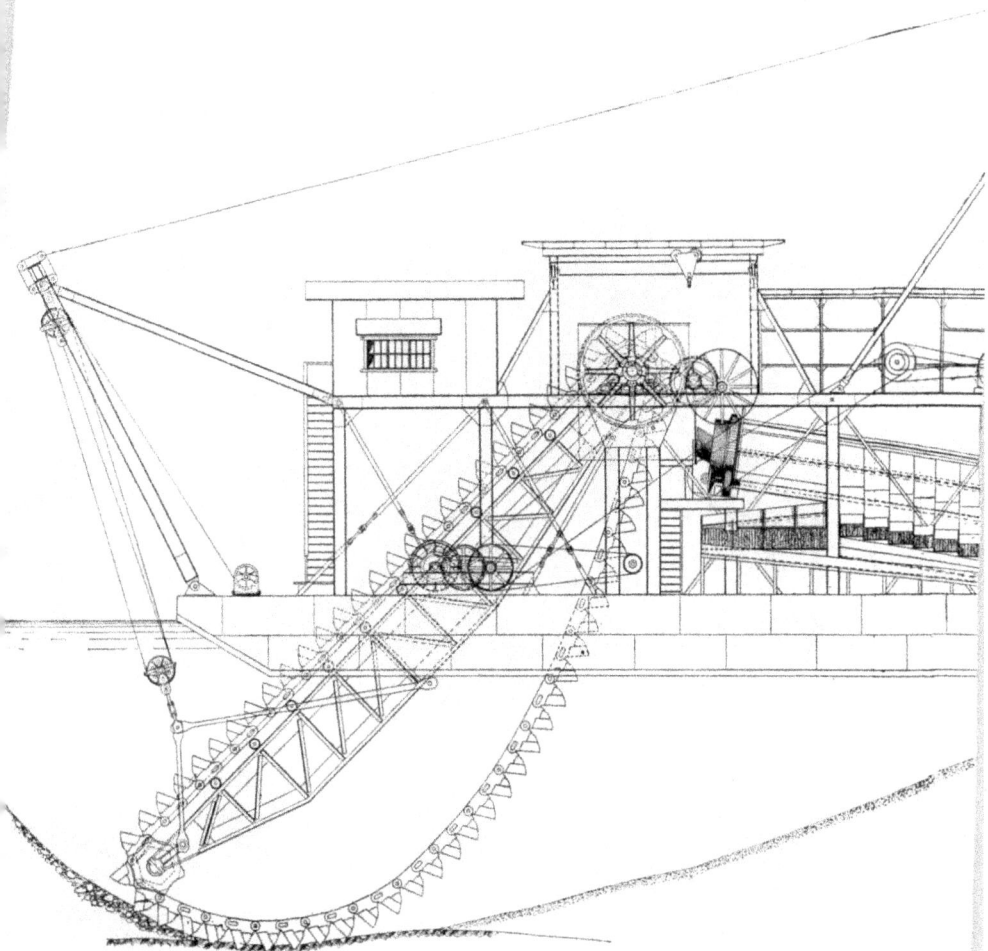

No. 63. Dredge of the Oroville Dredgin

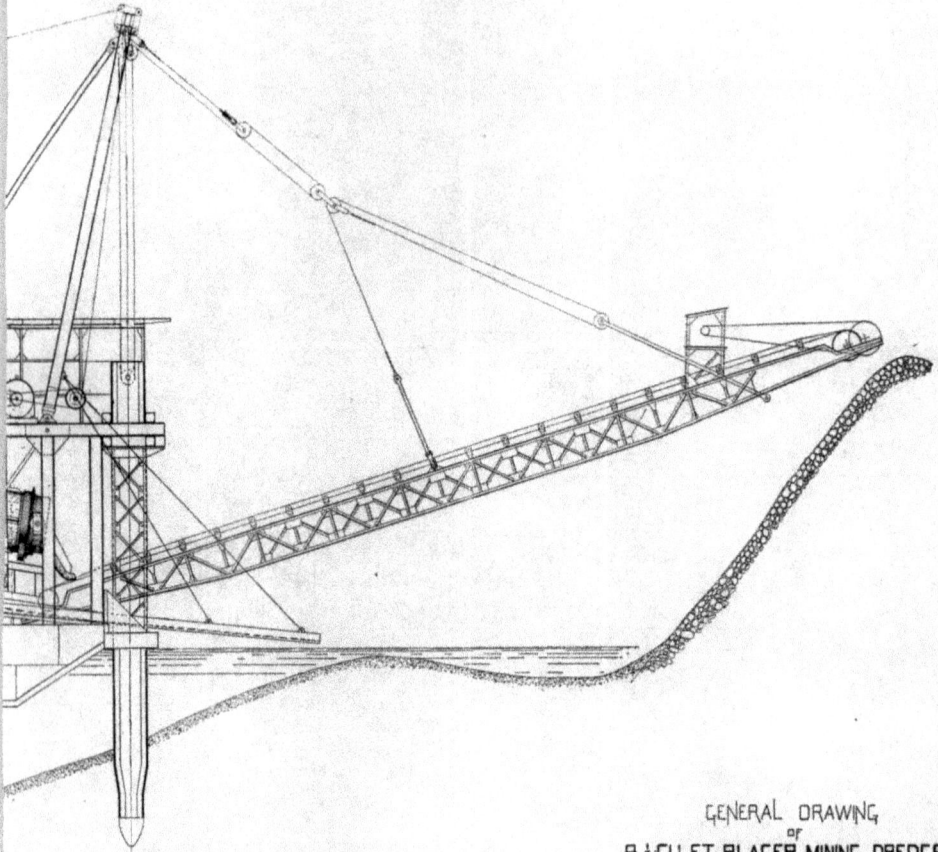

GENERAL DRAWING
OF
8½ CU. FT. PLACER MINING DREDGE
WITH
STEEL HULL
DESIGNED BY THE ENGINEERS
OF THE
YUBA CONSTRUCTION CO
MARYSVILLE, CAL. U.S

Co., Limited, Colombia, South America.

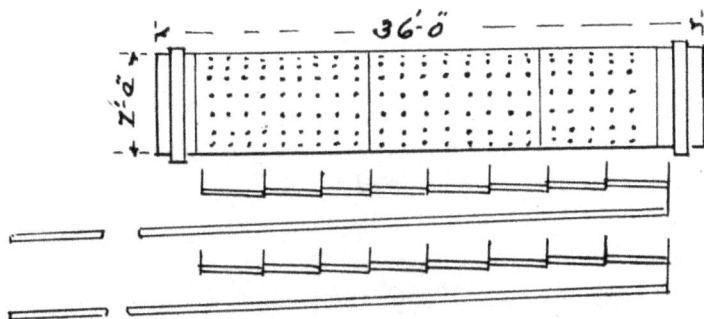

— ELEVATION —

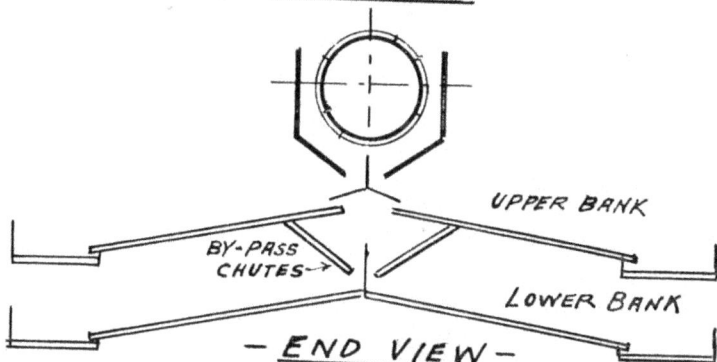

UPPER BANK

BY-PASS
CHUTES→

LOWER BANK

— END VIEW —

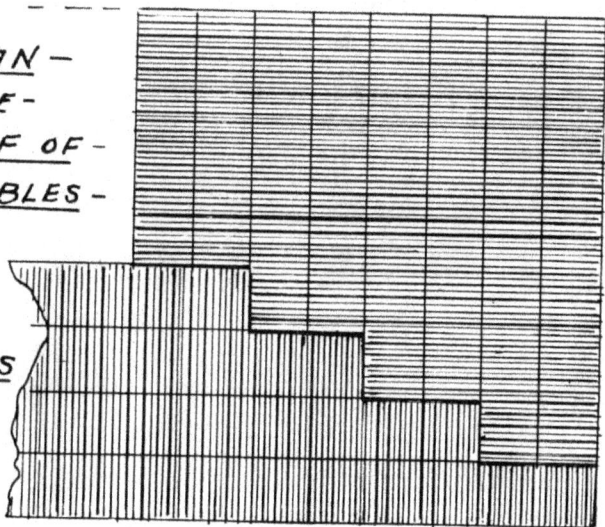

—PLAN—
- OF -
-HALF OF -
- TABLES -

SIDE
SLUICES

— PLAN —

No. 64. Double-bank gold-saving tables.

No. 65.   Lower bank of gold-saving tables on a double-bank dredge.   Folsom No. 6.

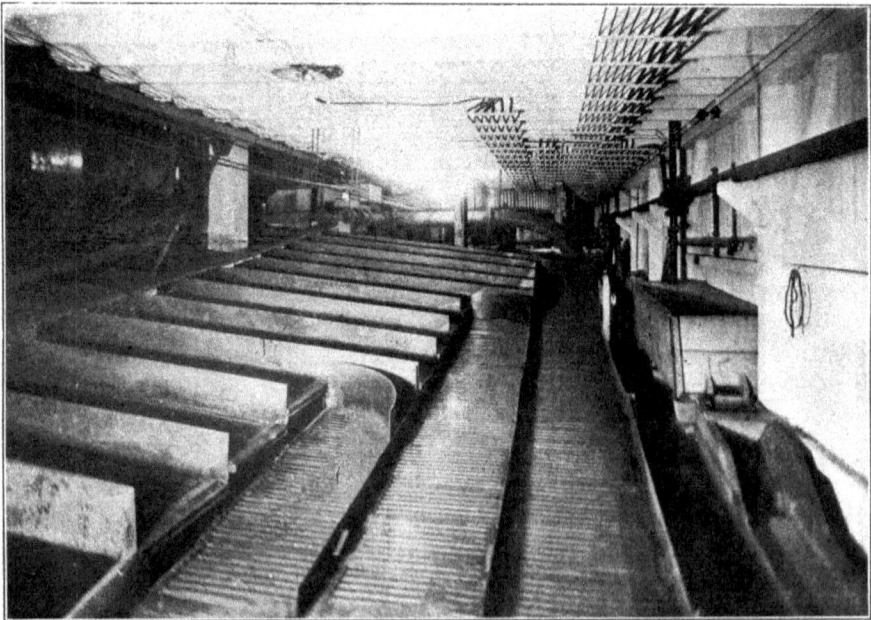

No. 66.   Upper bank of gold-saving tables on a double-bank dredge.   Clean-up box on right.
Folsom No. 6.

Among the many different designs of riffles, the angle-iron type of Hungarian riffle is considered by many as the most effective. This consists of a series of angle-irons spaced 2½-inch center with the horizontal leg placed so that it points downstream. The water passing over the riffles forms an eddy under the projecting leg, which protects the quicksilver and facilitates the catching of the gold. The riffle frames are set with the riffles across the current and are sometimes alternated with plain iron bar riffle frames set in the direction of the flow. Mercury is usually fed at regular intervals into the top of the sluice. Small quan-

No. 67. A general view of the Holmes type of gold-saving tables. Showing clean-up device suspended in background. El Oro Dredge. Oroville District.

tities of quicksilver in the sluices and the break in the current, caused by the eddy of the riffles, help to secure amalgamation.

The most efficient grade for the tables and sluices is considered to be 1½ inches per foot. The quantity of water used varies considerably and depends a great deal upon the amount and character of the material washed, as well as on the personal judgment of the operator. There are, probably, few cases in which the exact quantity of water used has been determined. Some operators claim they do not use more than 100 inches in general operations at any time, while others claim to use an average of 300 inches.

It is aimed to so regulate the grade of the tables or sluices and the flow of the water, that all the gold will come in contact with the quicksilver.   Fine gold may be carried off by the water flowing over the riffles in too great a quantity or under too much speed.   On the other hand, if the grade is too flat or the volume of water low, an accumulation of sand may cover the mercury and amalgamation be hindered.   This is especially the case when the gravel contains a great deal of black sand, which, owing to its weight, lodges readily between the riffles.

In the Holmes method of riffle tables and launders, a wide steel tray is placed directly under the screen, sloping in the same direction, and fitted either with or without riffles.   This steel tray, which catches the screening, empties on to another tray sloping in the opposite direction, which in turn delivers the material on to a set of divided sluices fitted with riffles and sloping towards the stern on either side of the dredge. The whole gold-saving arrangement is generally constructed of steel.

A system of riffles, patented by James H. Leggett, is composed of 1½-inch angle-iron, so arranged in the sluices that the iron lies longitudinally with the flow.   The space between the angle-iron is filled with medium sized pebbles, which lodge naturally when the gravel is allowed to flow over the riffles, immediately after clean-ups, through openings in the upper end of the screens arranged for this purpose.

The entire sluice system, which is continuous and not side-distributing, is equipped with these self-paving angle-iron riffles and is somewhat similar to the Holmes tables.   These riffles admit of rapid clean-ups, requiring less than one hour.   The gold is caught and sinks into the spaces between the pebbles; the gathering of amalgam on top of the riffles and the consequent risk of the gold being picked up by undissolved clay and lost is avoided.

Quicksilver is distributed over the entire area after each clean-up and repaving, and small quantities are added daily.   It is found in the clean-ups that where the water reverses in flowing from one sluice to another, that the first half of the succeeding table contains the concentrated amalgam.   Careful tests, made over a period of three months, demonstrated a high recovery of the gold content by this system.

Some boats are so arranged that should a streak of clay or gravel containing much of such plastic material which would be liable to pick up any gold or amalgam it comes in contact with while passing over the riffles be encountered, the winchman can, without interfering with the working of the dredge, divert this material into separate sluices.

The time required to clean up the tables and sluices varies on the different dredges, and the frequency of the operation depends upon the judgment of the man in charge and the nature of the material handled. It is important that the clean-up is not made oftener than is necessary to prevent loss by overloading the riffles.   The speed with which the clean-up is accomplished is also an important item, as the dredges are usually shut down during the operation, though in some cases the dig-

No. 68.   General view of single-bank gold-saving tables as commonly used in California.

ging is not interrupted, as the entire flow of material from the screen is diverted to some of the tables while the others are being cleaned. It is customary on clean-up days to overhaul the machinery and make any minor repairs that are necessary.

The method used in cleaning up varies according to the idea of the different operators, but the principle is much the same. Commencing at the upper end, the riffle frames are taken up and washed and laid aside. A hose is used to assist in gathering the amalgam and con-

No. 69. Folsom No. 4 Dredge, showing clean-up vat and sluice.

centrates, which are then scraped into buckets, deposited in vats, and later carefully fed into a small sluice, usually on the order of a long-tom equipped with stops, mercury traps, and riffles; the amalgam and quicksilver is collected and strained and the amalgam is retorted and melted. In some districts a great deal of shot lead, iron nails, coins and ornaments are picked up by the dredges and lodge in the tables. This material, which all collects some amalgam, forms a second product, which is treated whenever a sufficient quantity accumulates and shipped in the form of a base bullion.

The loss of gold from the gold-saving tables and sluices is, probably,

No. 70. Gold-saving tables on the Indiana No. 1 Dredge, old design.

in most cases a great deal less than is ordinarily supposed. Many tests have been made in the various districts to determine the amount of these losses, and in most cases the loss was found to be small.

It is very difficult to thoroughly sample the dredge tailing and to determine the difference between the actual gold content of the ground and the recovery on the dredge. In one of the most thorough tests made for this purpose a small bucket elevator was used, the material being taken through perforations in the bottom of the tail sluice and passed to a small wooden sump in which the lower sprocket wheel of the elevator was mounted. The material was somewhat concentrated owing

No. 71.   Corner in retorting room at Natoma.

to the method of collection, but this could not be avoided. The speed of the elevator was arranged so that at the end of each shift approximately two cubic yards of material would be collected, which was then carefully rocked and panned.

All of these tests showed an appreciable loss both of quicksilver and amalgam by the dredge, but what the exact percentage of loss was as compared to the gold content of the yardage handled during the time of the test is difficult to state. It is estimated by those competent to judge that the average loss in dredging, while varying with different conditions, methods of operating, and efficiency of labor, does not exceed 10 per cent of the total gold content of the material handled, and on carefully operated boats probably not over 5 per cent, which would compare favorably with any other form of placer mining.

Several tests were made by the Viloro Syndicate, Limited, and other companies at Oroville, to determine the platinum content of the ground and the loss of platinum in dredging operations. The loss of platinum, while appreciable, was found insufficient to warrant the installation of suitable concentrating machinery to recover same.

The working time of a dredge is twenty-four hours, divided into three eight-hour shifts. Each dredge is usually operated by a crew of nine men, one dredgemaster and one shoreman, making a total of eleven men. Occasionally additional men are employed, usually on day shift, as special work may require. The usual dredge crew of one shift consists of a winchman and two oilers or deckmen. The prevailing wages, in California, are:

Dredgemasters, from $125 to $200 per month.

Winchmen, $3 to $4 per shift of eight hours.

Oilers or deckmen, $2.50 or $3 per shift of eight hours.

Shoremen or outside labor, from $2 to $2.75 per nine-hour shift.

Chinese, $1.50 to $1.75 per nine-hour shift.

Where there are a number of dredges a clean-up man, with one or more assistants, is employed in addition to the regular dredge crew; under his direction the gold on the tables of each dredge is removed at regular intervals, usually once a week. The clean-up man usually has charge of the retorting of the amalgam and the melting of the gold also, and because of this last duty is often known as the company's gold man.

The cost of a complete California type dredge varies considerably and is governed by the depth to be dredged below water line and the character of material to be handled. The following figures give a general idea of the cost of different size dredges installed complete and ready for operation: 3 and 3½ cubic feet, from $50,000 to $60,000; 5 and 5½ cubic feet, from $50,000 to $90,000; 7 and 7½ cubic feet, from $80,000 to $120,000; 8 and 8½ cubic feet, from $120,000 to $175,000; 9 and 9½ cubic feet, from $130,000 to $200,000; 12 and 13½ cubic feet, from $175,000 to $225,000.

Several placer mining dipper dredges, ranging in sizes of from 1¼ to 2½-yard bucket capacity, have been operated in the Oroville district at various times. Among these, the "Plutus," of 1¾-yard capacity, owned by John W. Ferris, was one of the first dredges in the district.

The washing machinery, gold-saving tables, etc.,

No. 72. Old type dipper dredge, equipped with double hoppers, screens and stackers. Dredge of the Yreka Creek Gold Dredging and Mining Company. Moved to Alaska.

6—GD

No. 73.   Plutus, 1¾-yard placer mining dipper dredge.   Oroville District.   Old Marion type.

were installed on the same hull with the dredge machinery, which was placed a little to one side of the center of the boat, in order to admit of the washing machinery and gold-saving apparatus being placed on the opposite side.

The hull was 46 feet wide, 55 feet long, and 6 feet 6 inches deep. The boom was 40 feet long, and the dredge was capable of digging 22 feet below the water-line and dumping 18 feet above the surface of the water. This dredge was finally dismantled, as the ground was merged with that of other companies.

An electrically driven 1¼-yard bucket capacity dipper dredge was installed on the Oroville Gold Dredging Company's ground, and the dredge was capable of digging 23 feet below the surface of the water. The size of the hull was 40 feet wide, 80 feet long, and 7 feet deep, having a 40-foot boom.

This dredge was equipped with a perforated shaking screen, about 6 feet wide and 18 feet long, placed in front at the side of the dredge, with sluicing tables running from beneath the screen to a main sluice box, which carried the screenings aft. A belt conveyor was arranged to carry the coarse material from the lower end of the screen to a central hopper placed at the foot of an 80-foot swinging stacker or tailing elevator ar-

No. 74. Oroville Dredging Company's Marion dipper dredge, now dismantled.

ranged on the rear of the boat. The stacker could be swung from one side to the other in an arc of about 140 degrees to deposit the tailing at various points at the rear of the dredge. The dredge had a capacity of about 1,800 yards in twenty-four hours with one shaking-screen, which

could have been increased, as the framework in front of the boat was so arranged that another shaking-screen could have been added on the opposite side.

A somewhat similar dredge of 1¼-yard bucket capacity, electrically driven, was installed on Wyman's Ravine, near Oroville, for the Garden Ranch Gold Dredging Company. The boom on this dredge was 30 feet long. This dredge also had a perforated shaking-screen 6 feet wide and about 20 feet long, placed on one side of the boat with gold-saving tables thereunder, emptying into a main sluice on board the dredge, which carried the tailing to the rear of the boat. On account of the rather shallow ground, about 18 feet deep, a single belt conveyor about 80 feet long arranged at the side, was ample to take care of the tailings. This dredge with only one screen showed a capacity of from 30,000 to 40,000 cubic yards per month, which might possibly have been increased 33⅓ per cent by having another screen installed on the opposite side.

The total cost of the dredge turned over ready for operation f. o. b. railroad at Marion, Ohio, was approximately $25,000.

The builders and some operators claim that with the following conditions there is a field for the dipper type of dredge:

First—Where the ground is somewhat shallow.

Second—Where the extent of ground is not sufficient to warrant a costly dredge.

Third—Where the material is of a somewhat rough character, boulders, stumps, etc.

Fourth—Where the ground is mixed with more or less clay, as the dipper will relieve itself, notwithstanding the adhesiveness of the clay.

The Marion Steam Shovel Company build these dipper type of placer mining dredges in the 1¼, 1½, and 2½-yard sizes, either steam or electrically driven, and arranged with a central screen or washing apparatus a little to the rear of the center of the boat with gold-saving tables on either side, on practically the same lines as now used on the elevator type of dredge, requiring only one screen and one stacker to take the full capacity of the digging end of the dredge.

The crew required to operate a dipper dredge running three shifts consists of one dredgemaster, three levermen, three cranemen, three oilers, and one laborer.

IV. WORKING COSTS.

Different dredging companies have various methods of segregating costs and there is often considerable variation in the methods employed. In compiling the records for this book the endeavor has been to arrive at the total cost within some degree of accuracy. Working costs, in general, depend a great deal on methods of bookkeeping and can not be used in comparison unless a uniform system is employed. Ground even in the same locality often varies to such an extent that different dredges

No. 75.   Placer dredge operating on Australian River.

of the same make and bucket capacity and operating under the same management show entirely different working costs. The difference in cost per cubic yard where one dredge only or where several dredges are operating under one management is not great, working conditions being the same and the management capable and economical in both instances.

It is apparent that should too large a yardage be reported as having been dredged at a certain total cost, that the cost per cubic yard figured from these results would be too low. Where dredges are operating in swift running streams as in New Zealand, and elsewhere, the measurement of yardage with any degree of accuracy is practically impossible. In such cases the yardage is usually computed by counting the buckets dumped per minute, some allowance being made for buckets only partly filled. When dredges are owned and operated by individuals the practice of keeping exhaustive records, both as to yardage handled and segregated items of working costs, is not generally followed.

In California all dredging companies arrive at the yardage handled

No. 76.   Monitors at work on cemented gravel bank.   California type dredge.

by measuring the gravel bank ahead of the dredge.   Careful daily records are in most instances kept of all costs.

While high yardage and low working costs are desirable, it is not the aim of the California dredge operator to make records at the expense of good work, but to strive for the best economic results, and it is with this end in view that the California type of dredge has been developed.

For many years small and light dredges, equipped with open connected buckets, head lines and short tray tailing stackers were used in California with varying success, but as the loose shallow ground was worked out and the gravel became deep and at times cemented, the large California type of dredge equipped with heavy machinery, spuds, and belt tailing conveyor proved its superiority.

No. 77. Showing ladder and bucket line on 5-cubic-foot dredge, Oroville, Cal.

3-CUBIC-FOOT CLOSE-CONNECTED-BUCKET ELEVATOR DREDGE.

Remodeled from open-connected. Five years and nine months in commission. New Zealand type.

Summary of working costs for the last twelve months in operation.

Actual dredging time, 7,216 hours 40 minutes; average dredging time daily, 19 hours 43 minutes; cubic yards excavated, 458,882; average yardage daily, 1,254; acres dredged, 10.57; average depth of ground, 26.9 feet.

	Total Costs.	Cost per Cubic Yard in Cents.	Per Cent of Total Cost.
Operating, material and labor	$9,325 30	2.03	28.9
Electricity	3,170 49	.69	9.9
Repairs	15,071 92	3.28	46.9
General expense	2,913 25	.63	9.1
Taxes and insurance	1,685 88	.37	5.2
Total expense	$32,166 84	7.00	100.

This dredge was working under very favorable conditions in rich ground.

3-CUBIC-FOOT CLOSE-CONNECTED-BUCKET ELEVATOR DREDGE.
Remodeled from open-connected.   New Zealand type.
Working costs for the last twelve months the dredge was in operation.

Actual dredging time, 2,809 hours 30 minutes; average dredging time daily, 16 hours 3 minutes; cubic yards excavated, 173,665; average yardage daily, 992; acres dredged, 3.97; average depth of ground, 27.1 feet.

	Total Costs.	Cost per Cubic Yard in Cents.	Per Cent of Total Cost.
Operating, material and labor	$4,816 69	2.77	30.1
Electricity	1,565 40	.90	9.8
Water	240 00	.14	1.4
Repairs	7,199 63	4.15	44.9
General expense	1,358 43	.78	8.5
Taxes and insurance	845 26	.49	5.3
Total expense	$16,025 41	9.23	100.

Gross returns per cubic yard, 7.66 cents; loss, 1.57 cents.

This dredge was working on a headline in difficult ground where now a new 7-cubic-foot spud dredge is working at a profit.

3½-CUBIC-FOOT CLOSE-CONNECTED-BUCKET ELEVATOR DREDGE.
Six and one half years in commission.   California type.
Working costs for the last twelve months in operation.

Actual dredging time, 305 days 21 hours; average dredging time daily, 20 hours 30 minutes; cubic yards excavated, 461,882; average yardage daily, 1,510; acres dredged, 7.4, average depth of ground, 34 feet.

	Cost per Cubic Yard in Cents.
Operating, labor	2.853
Power, electric	1.487
Water	.195
Repairs, material and labor	1.717
General expenses, taxes, insurance, smelting and express	1.077
Total expenses	7.329

Gross output, $73,475.91; net output, $39,628.37.

Operating under difficult conditions, owing to location of land, which is subject to overflow during high water.   The gravel is a fairly compact clean river wash.

3½-CUBIC-FOOT CLOSE-CONNECTED-BUCKET ELEVATOR DREDGE.
California type.
Working costs during the first seven years in operation.

	1st Year.	2d Year.	3d Year	4th Year.	5th Year.	6th Year.	7th Year.
Labor	2.913	3.133	3.975	2.687	3.011	2.853	2.838
Power	1.886	1.960	2.467	1.542	1.446	1.487	1.535
Water		0.180	0.255	0.194	0.211	0.195	0.228
Repairs and supplies	2.415	3.383	2.624	2.515	2.395	1.717	1.743
General expenses	0.745	0.658	0.797	0.672	1.354	1.077	1.323
Total costs, in cents	7.959	9.314	10.118	7.610	8.417	7.329	7.667
Cubic yards handled	485,016	466,262	352,826	465,207	425,843	461,882	395,316

In the above costs are included the operating expenses, total cost of repairs necessary to keep the dredge in the best working condition, also the cost of all extraordinary breakages and accidents.

The dredge was digging to an average depth of about 35 feet in fairly compact clean gravel, carrying little clay and few large boulders. The ground is subject to overflow during high water, thus increasing the working costs during the winter months.

4-CUBIC-FOOT CLOSE-CONNECTED-BUCKET ELEVATOR DREDGE.

Nine years in commission.   Early Montana type remodeled to California type.

Working costs for the last twelve months in operation.

Actual dredging time, 7,057 hours; average dredging time daily, 19 hours 20 minutes; cubic yards excavated, 484,387; average yardage daily, 1,327; acres dredged, 14.60; average depth of ground, 20.6 feet.

	Total Cost.	Cost per Cubic Yard in Cents.	Per Cent of Total Cost.
Operating, material and labor	$8,832 08	1.83	27.9
Electricity	4,330 04	.89	13.7
Water	1,500 00	.31	4.8
Repairs	12,498 00	2.58	39.5
General expense	3,173 96	.65	10.0
Taxes and insurance	1,266 85	.26	4.1
	$31,600 93	6.52	100.

Digging on low land subject to overflow in flood times. The gravel is mostly a clean river wash, in places very shallow. The character of the bedrock is a soft tufa with uneven contours. This dredge was remodeled from a double-lift to a single-lift.

5-CUBIC-FOOT CLOSE-CONNECTED-BUCKET ELEVATOR DREDGE.

Four and one half years in commission.   California type.

Working costs for the last twelve months in operation.

Actual dredging time, 6,507½ hours; acres dredged, 13.96; cubic yards excavated, 812,355; average depth of ground, 36.06 feet.

	Cost per Cubic Yard in Cents.	Per Cent of Total Cost.
Operating,* material and labor	2.301	34.568
Power, electric	1.081	16.231
Repairs, material and labor	2.923	43.890
Taxes and insurance	0.354	5.311
	6.659	100.

* Under "Operating" is included all expenses for salaries, bullion, general expenses, labor and supplies, and water.

Digging in average ground, carrying considerable clay. The surface of the ground is covered by a thick growth of small timber which must be cleared before dredging.

5-CUBIC-FOOT CLOSE-CONNECTED-BUCKET ELEVATOR DREDGE.

Three years and five months in commission. California type.

Summary of working costs for the last twelve months in operation.

Actual dredging time, 6,798 hours 45 minutes; average dredging time daily, 18 hours 35 minutes; cubic yards excavated, 1,148,480; average yardage daily, 3,138; acres dredged, 27.91; average depth of ground, 25.5 feet.

	Total Costs.	Cost per Cubic Yard in Cents.	Per Cent of Total Cost.
Operating, material and labor	$10,066 19	.88	25.1
Electricity	6,031 50	.52	13.8
Water	603 34	.05	1.4
Repairs	20,271 98	1.77	46.4
General expense	2,913 26	.25	6.7
Taxes and insurance	3,766 93	.33	8.6
Total expense	$43,653 20	3.80	100.

This dredge is working in loose gravel with a heavy overburden of sandy loam.

5-CUBIC-FOOT CLOSE-CONNECTED-BUCKET ELEVATOR DREDGE.

Two years and five months in commission. California type.

Working costs for the last twelve months in operation.

Actual dredging time, 6,790 hours 35 minutes; average dredging time daily, 18 hours 36 minutes; cubic yards excavated, 1,148,802; average yardage daily, 3,148; acres dredged, 23.87; average depth of ground, 29.9 feet.

	Total Costs.	Cost per Cubic Yard in Cents.	Per Cent of Total Cost.
Operating, material and labor	$9,475 59	.82	22.7
Electricity	5,607 00	.49	13.4
Water	300 00	.03	0.7
Repairs	21,719 28	1.89	52.0
General expense	2,865 88	.25	6.9
Taxes and insurance	1,820 10	.16	4.3
	$41,787 85	3.64	100.

No. 78. Ground to be dredged, partly cleared of brush and trees. Oroville District.

This dredge was digging in loose gravel with a heavy overburden of light soil. The heavy cost of maintenance is due mostly to the installing of a new upper tumbler, a complete line of bucket bottoms, and a new stacker belt.

5-CUBIC-FOOT CLOSE-CONNECTED-BUCKET ELEVATOR DREDGE.

Four years and seven months in commission. California type.

Working costs for the last twelve months in operation.

Actual dredging time, 6,644 hours 20 minutes; average dredging time daily, 18 hours 12 minutes; cubic yards excavated, 599,614; average yardage daily, 1,643; acres dredged, 9.66; average depth of ground, 38.5 feet.

	Total Costs.	Cost per Cubic Yard in Cents.	Per Cent of Total Cost.
Operating, material and labor	$10,599 83	1.77	23.1
Electricity	5,490 00	.92	11.9
Water	1,500 00	.25	3.3
Repairs	24,127 64	4.03	52.2
General expense	2,836 61	.47	6.1
Taxes and insurance	1,402 75	.23	3.1
Total expense	$45,956 83	7.67	100.

Working under great difficulties against a 20-foot bank and rising bedrock. It was necessary to cut through the bedrock, which rose above the water level to a sufficient depth to enable the dredge to maintain its course.

Heavy repair cost, due to installing of new tumbler, conveyor belt, repairs to digging ladder, and replacing of screen, etc., accounts for 52.2 per cent of the total working cost.

Yuba Consolidated Gold Fields' dredges, each of 6-cubic-foot capacity, California type dredges.

5-CUBIC-FOOT CLOSE-CONNECTED-BUCKET ELEVATOR DREDGE.

Four and one half years in commission.

Working costs, labor and material only, for the last twelve months in operation.

Actual dredging time, 6,507 hours 20 minutes; acres dredged, 13.96; cubic yards excavated, 812,355; averaged depth of ground 36.6 feet.

REPAIR COSTS IN LABOR AND MATERIAL.

	1907		1908										Totals			
	Nov.	Dec.	Jan.	Feb.	March.	April.	May.	June	July.	Aug.	Sept.	Oct.	Labor.	Material.	Total.	Per Cent.
Ladder—																
Labor	$16 11	$45 61	$73 47	$74 27	$2 75	$31 75		$431 82	$18 01	$82 45	$27 69	$74 99	$978 92		$2,079 65	8.760
Material	77 76	78 79	95 39	5 50	39 08	21 32	$75 45	473 06	6 60	177 85	3 00	26 03		$1,100 73		
Bucket line—																
Labor	60 30	136 69	90 56	106 12	71 87	37 67	21 81	14 63	4 59	10 75	16 46	1 86	573 31		8,587 13	36.169
Material	709 00	780 73	722 45	768 68	733 50	761 46	729 00	709 00	700 00	700 00		700 00		8,013 82		
Tumblers—																
Labor	113 95	50 50	64 32	33 05	37 60	61 35	113 37	87 04	82 31	8 93	22 46	11 79	686 87		3,017 87	12.711
Material	571 80	2 86	65 11	61 96	399 19	282 06	27 73	191 16	226 44	193 35	21 72	287 62		2,331 00		
Screens—																
Labor	43 78	8 58	11 84	5 49	6 99	39 46	90	14 33	6 03	18 17	29 37	48 43	233 28		1,104 11	4.651
Material	38 33	45 33		88	91 51	92 54	25 32	2 45	71 74	291 58	292 39	8 76		870 83		
Spuds—																
Labor	38 98		57 16	9 25	1 44	3 10	24 96	1 50	9 09	24 94	5 75	64 73	239 40		337 34	1.589
Material	4 50		116 77				15 17							137 94		
Stacker—																
Labor	48 14	8 09	20 86	98 04	1 04	6 08	17 25	9 53	22 62	17 23	53 51	47 02	391 79		2,392 92	10.079
Material	231 41	27 10	45 18	665 20	658 85	1 30	14 80			24 80	375 27			2,091 13		
Pumps—																
Labor	10 79	88 14	28 22	1 17	24 79		1 32	43 41	25 59		8 94		233 14		641 80	2.703
Material	18 12	35 98	31 06	5 41	56 15			18 94	243 00					408 66		
Winches—																
Labor	37 31	35 73	27 92	62 22	43 82	14 58	17 78	21 29	35 74	20 56	20 23	61 02	398 20		1,416 69	5.967
Material		50 33	30 84	121 88	193 80	16 39	129 75	14 34	31 50	243 45	3 40	182 81		1,018 49		
Lines—																
Labor	11 88	8 38	19 93	4 41	1 46	15 36	7 11	5 39	21 77	32 17	3 40	25 16	156 42		1,817 50	7.655
Material	4 50	465 42	248 66	3 42	127 30	45 54	377 62	2 10		266 30	1 50	118 82		1,661 08		
Motors—																
Labor					57					6 39			6 96		394 53	1.662
Material	21 58	49 60	6 50	15 26	13 75	50 64		1 95	23 63	113 65	84 48	6 53		387 57		
Other causes—																
Labor	77 13	129 11	73 87	76 41	40 68	80 08	69 35	36 04	115 65	64 83	86 54	80 16	929 85		1,912 24	8.054
Material	105 39	52 37	67 08	68 73	15 90	75 81	100 15	119 04	39 36	141 43	51 33	86 50		982 39		
Totals	$2,340 76	$2,099 34	$1,807 29	$2,187 35	$2,552 15	$1,636 69	$1,829 04	$2,197 02	$1,713 07	$2,439 40	$1,017 44	$1,882 23	$4,738 14	$19,003 64	$23,741 78	100.

OPERATING COSTS IN LABOR AND MATERIAL.

	1	2	3	4	5	6	7	8	9	10	11	12	Labor	Material	Total	Per cent
Operating—Labor	$533 10	$649 11	$651 64	$585 43	$472 28	$619 64	$829 50	$462 09	$967 68	$975 76	$924 97	$849 49	$7,290 79	$1,014 48	$8,305 27	75.120
Material	159 68	71 91	291 64	12 89	25 11	10 50		75 36	197 55	50 31	102 18	17 35				
Power line—Labor	3 33		8 50	1 50	24 87	1 56	2 94	4 71		7 46			55 79		78 14	.707
Material					22 35							92		22 35		
Clean-ups—Labor	11 40	16 55	11 34	12 21	17 53	20 21	17 43	17 13	16 82	18 75	16 82	21 89	198 08		198 08	1.792
Material																
Other causes—Labor				18 50	242 97	19 91				25 55			306 93	135 42	442 35	4.001
Material				36 79	83 63	15 00										
Clearing—Labor	129 49	116 11	96 12	66 36	132 12	91 54	98 86	56 87	84 86	76 99	81 87	81 61	1,115 80	916 33	2,032 13	18.389
Material	55	53 00	34 10	32 75	186 03	432 35	12 00	31 15	81 33	24 82		28 25				
Totals	$837 55	$906 68	$1,093 34	$716 43	$1,406 89	$1,210 71	$760 73	$647 31	$1,048 34	$879 64	$828 84	$719 51	$8,967 39	$2,088 58	$11,055 97	100.
Repair totals	2,340 76	2,099 34	1,897 29	2,187 35	2,552 15	1,636 69	1,829 04	2,197 02	1,713 07	2,439 40	1,017 44	1,832 23	4,738 14	19,003 64	23,741 78	
Grand totals	$3,178 31	$3,006 02	$2,990 63	$2,903 78	$3,959 04	$2,847 40	$2,589 77	$2,844 33	$2,761 41	$3,319 04	$1,846 28	$2,551 74	$13,705 53	$21,092 22	$34,797 75	
Per cent	9.134	8.638	8.594	8.345	11.377	8.183	7.442	8.174	7.935	9.539	5.306	7.333	39.386	60.614	100.	

6-CUBIC-FOOT CLOSE-CONNECTED-BUCKET ELEVATOR DREDGE.

Four years in commission.

Summary of working costs for the last twelve months in operation.

Working days in year	363
Working hours in year	8,712
Possible working hours	8,601½
Total hours machine was digging during year	5,730 5-6
Power was off, hours	110½
Percentage of total hours digging	65.42
Percentage of possible hours digging	66.62
Surface area worked out, acres	11.956
Average depth of cut, feet	25.12
Cubic yards handled	541,744
Cubic yards handled per day	1,492
Cubic yards handled per digging hour	94.52
Gross value extracted	$61,646 37
Gross value extracted per day	$169.82
Gross value extracted per digging hour	10.75
Gross value extracted per yard dug	.11380

Division of costs.

	Cost per Cubic Yard Dug.	Total Expense.
Labor	$0.02430	$13,163 68
Power	.01328	7,196 61
Supplies	.00703	3,810 47
Shop repairs and parts	.01055	5,718 09
Freight	.00118	638 57
General expenses and boarding house, loss	.00130	706 30
Land clearing	.00045	244 10
Total cost at dredge	$0.05809	$31,477 82
Profit over and above cost at dredge	$0.05571	$30,168 55

Other costs—Administration, taxes, insurance, etc.

San Francisco office, management and traveling expenses	$0.00759	$4,116 15
Treasurer's office expenses	.00041	225 00
Taxes	.00180	922 67
Insurance	.00086	521 63
Total administration	$0.01066	$5,785 45
Add costs at dredge as above	.05809	31,477 82
Total costs all included	$0.06875	$37,263 27

_____ DREDGING COMPANY

Operations by Years

No. 79. Yearly record sheet, used by some Ca[...]

_____ DREDG[...]

Time Sheet for the Month of ___ *February* ___

	TIME LOST									
	LADDER AND BUCKET LINE	STACKER	WINCHES	SCREEN	WATER PUMP	SAND PUMP	LINES	POWER	OTHER CAUSES	TOTAL LOST
Date	Hrs Mins	Hrs Mins	Hrs Mins	Hrs Mins	Hrs Mins	Hrs Mins	Hrs Mins	Hrs Mins	Hrs Mins	Hrs Mins
1	25	30		10						1 05
2							20		1 00	1 20
3		20		10						30
4				20			20		10	50
5		1 05		10						1 15
6		2 40								2 40
7		25		10						35
8				10			10		10	30
9	10			10			10		10	40
10				20						20
11		1 15		10				25	10	2 00
12				10			1 20		10	1 40
13	3 55	05								4 00
14	3 20	10					10		3 05	6 45
15	5 00	05								5 05
16	1 25			35			10			2 10
17	05			10			10		10	35
18		30		10			30			1 20
19		45		10			10			1 05
20		05		15			40		3 15	4 15
21	10			40					10	1 00
22		30		30			2 00			3 00
23		10					20	10		40
24		20				35	10			1 05
25				20					05	25
26		05		15			25		1 20	2 05
27	0 5	10					10		10	35
28		25							15	40
29										
30										
31								•		
Total	14 35	9 35		5 15		35	7 15	25	10 20	48 10
Per Cause	2.13	1.42		.78		.09	1:08	.09	1:54	7.17 / 23.07

Sand Pump ran 201 Hrs 10 Mins.   50183 Cubic yards dredged
Water — 0 - 0 —   1792 Cubic yards per [...]
1930 — — dre[...]

No. 80. Monthly Time Sheet as used by some [...]

...GING COMPANY

_____19.0

TIME OPERATING					REMARKS
STEPPING AHEAD	CLEAN UP	DREDGING	TOTAL OPERATING	logs	
Hrs. Mins.	Hrs. Mins.	Hrs. Mins.	Hrs. Mins.		
1 00		21 55	22 55	31	
40	4 00	18 00	22 40	31	
50		22 40	23 30	31	
50		22 20	23 10	31	
50		21 55	22 45	31	
50		20 30	21 20	31	Put pad on stacker drive drum 225
50		22 35	23 25	31	
1 10		22 20	23 30	31	
40		22 40	23 20	31	
50		22 50	23 40	31	
40		21 20	22 00	31	Laced stacker Belt .50
50		21 30	22 20	31	Moved Stb lines 1.20
40		19 20	20 00	31	Changed buckets ebushings 305 Bottom ebr tumblers ladder 335
50		16 25	17 15	31	Turning Boat 225
30		18 25	18 55	35	Replaced Bar in Dump Hopper 405
30		21 20	21 50	35	Apron eBars in Dump Hopper 125
20		23 05	23 25	35	
30		22 10	22 40	35	
40		22 15	22 55	35	
40		19 05	19 45	35	Removing Broken Landing Bridge
20		22 40	23 00	35	
20		20 40	21 00	35	Spliced Bow Line 120
30		22 50	23 20	35	
20		22 35	22 55	35	
30		23 05	23 35	35	
20		21 35	21 55	35	Broke deadman Strap 120
40		22 45	23 25	35	
30		22 50	23 20	35	
18 10	4 00	601 40	623 50		
			262 00		
27 0	40	69 53	92 83		

Total hours in Month 672

Per cent of time operating 92.83

" " " " idle 7.17

7-CUBIC-FOOT CLOSE-CONNECTED-BUCKET ELEVATOR DREDGE.

California type.

Working cost during part of the second and third years in operation.

Month Ending.	Dredging Time. Hrs.	Mins.	Cubic Yards Excavated.	Average Depth, Feet.	Current Expenses. Labor and Material.	Electric Power.	Water.	Repairs.	General Expense.	Taxes and Insurance.	Total Expense.	Total Cost Cubic Yard, Cents.
August 31	511	20	74,687	37.6	$872.83	$663.73	$40.00	$1,311.65	$209.57		$3,200.80	4.29
September 31	480	50	56,052	27.9	904.11	632.25	40.00	800.39	163.29		2,540.24	4.53
October 31	568	00	65,835	35.7	987.45	670.50	40.00	2,236.68	374.08		4,308.71	6.54
November 30	523	20	64,010	27.0	1,028.18	641.25	40.00	4,337.15	169.73		6,216.31	9.71
December 31	478	20	76,015	31.0	1,069.31	540.00	40.00	4,251.74	151.54		6,052.59	7.96
January 31	584	45	106,913	32.7	1,029.85	789.50	40.00	2,391.04	327.07		4,557.46	4.26
February 29	561	40	101,377	35.6	963.55	767.25	50.00	1,425.56	145.14		3,351.50	3.31
March 31	638	50	101,110	34.7	1,497.51	756.00	50.00	2,191.00	201.85		4,296.36	4.25
April 30	571	55	84,631	32.6	1,080.52	717.75	63.33	2,130.34	477.98		4,419.92	5.22
May 31	624	45	92,763	33.7	1,073.79	740.25	66.68	2,303.18	162.69		4,346.59	4.68
June 30	589	50	86,946	32.1	1,052.11	780.25	66.68	2,304.63	240.44		4,444.86	5.11
July 31	189	20	24,963	33.9	598.37	213.73	66.68	2,896.13	289.93		4,064.86	16.28
										$3,235.45	3,235.45	.34
Totals	6,352	55	935,332		$11,807.61	$7,893.00	$663.37	$28,582.69	$2,913.31	$3,235.45	$55,035.43	
Averages	Daily 17	Daily 21	Daily 2,555	Actual. 33.4	1.26	.85	.06	3.06	.31	.34		5.88
Cost per cubic yard, in cents					21.15	11.3	1.1	51.9	5.3	5.9		5.88
Percentage of total cost					21.15	11.3	1.1	51.9	5.3	5.9		100.

During the month of July the dredge was idle for a period of three weeks undergoing repairs, which accounts for the high cost during that month.

7-CUBIC-FOOT OPEN-CONNECTED-BUCKET ELEVATOR DREDGE.

Two years in commission. Improved New Zealand type.

Summary of working costs for the last twelve months in operation.

OPERATING EXPENSES.

Month	Labor.	Power.	Water.	Mainte nance.	Total.	Cubic Yards Han dled.	Cost per Yard, Cents.	Per Cent Time Run ning.	Value of Clean-up.	Value per Yard, Cents.
July	$1,329 43	$555 00	$250 00	$1,152 17	$3,286 60	52,270	6.3	77	$8,244 18	15.7
August	923 00	357 00	250 00	1,030 76	2,560 76	35,200	7.3	50	5,171 39	14.8
September	1,320 95	531 30	250 00	856 57	2,958 82	46,700	6.3	68	4,345 23	9.3
October	1,200 00	492 00	250 00	1,178 30	3,120 30				5,546 44	
November	1,249 61	535 80	250 00	995 59	3,031 00	46,500	6.5	71	7,411 82	16.0
December	941 22	484 80	250 00	1,281 09	2,957 11	44,000	6.8	69	7,801 60	17.7
1908—January	1,245 20	514 20	250 00	1,270 52	3,309 92	44,050	7.5	72	6,529 42	14.8
February	897 94	558 00	250 00	1,334 68	3,040 62	38,800	7.6	60	6,394 75	16.48
March	1,300 00	584 40	250 00	1,131 50	3,265 90	40,900	7.1	65	7,399 74	18.4
April	880 81	475 20	250 00	1,272 33	2,878 34	37,600	7.6	63	5,265 39	14.
May	1,288 54	645 00	250 00	1,257 35	3,440 89	56,300	7.1	84	4,749 37	8.5
June	1,267 27	608 40	250 00	1,125 69	3,251 36	48,200	7.6	70	5,479 48	11.35
Totals	$13,843 97	$6,371 10	$3,000 00	$13,886 55	$37,101 62	490,520	7.06		$74,338 81	14.02

EXTRAORDINARY EXPENSES.

Month	Risdon.	Power Off.	Lost Time.	Pennsylv'a Steel Co.	Sundries.	Total.
July					$247 74	$247 74
August	$1,764 32		$585 53		131 67	2,481 52
September		$429 00				429 00
October			241 69		252 92	494 61
November		150 00			59 75	209 75
December	228 81	525 00		$766 48	163 75	1,684 04
1908—January	952 87	175 00			491 67	1,619 54
February		100 00	300 00		261 70	661 70
March	257 20	75 66		drill lit	814 53	1,147 39
April	559 95	50 00	450 00		779 24	1,839 19
May	314 03			1,349 42	111 11	1,774 56
June	627 05			400 81	415 44	1,443 30
Totals	$4,704 23	$1,504 66	$1,577 22	$2,516 71	$3,729 52	$14,032 34

Buckets in pond.

Timbering, breakage and repairing.

NOTE.—The running time averages about 70 per cent, handling 525,000 cubic yards; if the dredge had run full time 750,000 cubic yards would have been handled. The difference, therefore, of 225,000 cubic yards at 14½ cents per yard would make a principal of $32,625, which at 6 per cent would equal $1,957.50 interest, which would not have accrued, since the $32,625 would have cancelled the original principal to that extent.

This dredge is digging in partly cemented gravel, averaging in depth about 35 feet.

7-CUBIC-FOOT CLOSE-CONNECTED-BUCKET ELEVATOR DREDGE.

Three years and nine months in commission. California type.

Summary of working costs for the last twelve months in operation.

Actual dredging time, 6,390 hours 20 minutes; average dredging time daily, 17 hours 28 minutes; cubic yards excavated, 1,033,694; average yardage daily, 2,824; acres dredged, 24.23; average depth of ground, 26.5 feet.

	Total Costs.	Cost per Cubic Yard in Cents	Per Cent of Total Cost.
Operating, material and labor	$11,155 55	1.08	21.2
Electricity	6,640 80	.64	12.6
Water	1,500 00	.14	2.9
Repairs	27,786 14	2.69	52.8
General expense	3,483 78	.34	6.6
Taxes and insurance	2,027 66	.20	3.9
Total expense	$52,593 93	5.09	100.

Gross returns per cubic yard, 11.02 cents; net returns per cubic yard, 5.93 cents.

This dredge is working in heavy compact gravel. The replacing of the upper and lower tumbler shafts, new revolving screen, and conveyor belt accounts for the loss in working time and heavy repair costs.

7-CUBIC-FOOT CLOSE-CONNECTED-BUCKET ELEVATOR DREDGE.

One year and three months in commission. California type.

Working costs for the last twelve months in operation.

Actual dredging time, 6,313 hours 35 minutes; average dredging time daily, 17 hours 18 minutes; cubic yards excavated, 1,114,605; average yardage daily, 3,054; acres dredged, 25.05; average depth of ground, 27.6 feet.

	Total Costs.	Cost per Cubic Yard in Cents.	Per Cent of Total Cost.
Operating, material and labor	$13,467 22	1.21	29.7
Electricity	6,954 27	.62	15.3
Water	315 00	.03	0.7
Repairs	20,192 95	1.81	44.5
General expense	3,173 96	.29	7.
Taxes and insurance	1,266 85	.11	2.8
Total expense	$45,370 25	4.07	100.

Gross returns per cubic yard, 9.76 cents; net revenue per cubic yard, 5.69 cents.

Digging in compact gravel.

7-CUBIC-FOOT CLOSE-CONNECTED-BUCKET ELEVATOR DREDGE.

Two years and nine months in commission. California type.

Working costs for the last twelve months in operation.

Actual dredging time, 6,917 hours 25 minutes; average dredging time daily, 18 hours 57 minutes; cubic yards excavated, 1,017,167; average yardage daily, 2,787; acres dredged, 22.43; average depth of ground, 28.1 feet.

	Total Costs.	Cost per Cubic Yard in Cents.	Per Cent of Total Cost.
Operating, material and labor	$11,188 09	1.10	24.4
Electricity	6,603 98	.65	14.4
Water	1,500 00	.15	3.3
Repairs	22,330 72	2.19	48.7
General expense	2,836 67	.28	6.2
Taxes and insurance	1,402 74	.14	3.0
Total expense	$45,862 20	4.51	100.

Gross returns per cubic yard, 19.11 cents; net returns per cubic yard, 14.60 cents.

Working under difficulties in heavy, compact ground. In the cost of repairs is included the cost of a new steel spud and a new revolving screen. The time lost is due to the breaking of the steel spud.

7-CUBIC-FOOT CLOSE-CONNECTED-BUCKET ELEVATOR DREDGE.

Nine months and ten days in commission. California type.

Actual dredging time, 5,088 hours 25 minutes; average dredging time daily, 17 hours 52 minutes; cubic yards excavated, 838,885; average yardage daily, 2,943; acres dredged, 14.66; average depth of ground, 35.5 feet.

	Total Costs.	Cost per Cubic Yard in Cents.	Per Cent of Total Cost.
Operating, labor and material	$9,980 60	1.19	33.7
Electric power	5,805 00	.69	19.6
Water			
Repairs	10,200 97	1.22	34.4
General expense	2,217 73	.26	7.5
Taxes and insurance	1,421 19	.17	4.8
Total expense	$29,625 49	3.53	100.

Gross returns per cubic yard, 6.36 cents; net returns per cubic yard, 2.83 cents.

Digging in heavy, difficult ground; gravel coarse, compact and partly cemented. This dredge replaced an open-connected-bucket dredge, which was working on this property and could not handle the ground to a profit.

7½-CUBIC-FOOT CLOSE-CONNECTED-BUCKET ELEVATOR DREDGE.

Nine months and six days in commission. California type.

Summary of working costs.

Actual dredging time, 5,582 hours 15 minutes; average dredging time daily, 19 hours 52 minutes; cubic yards excavated, 944,879; average yardage daily, 3,363; acres dredged, 20.425; average depth of ground, 28.7 feet.

	Total Costs.	Cost per Cubic Yard in Cents.	Per Cent of Total Cost
Operating, labor and material	$8,939 64	.95	26.7
Electricity	5,440 80	.58	16.3
Repairs	12,308 73	1.30	36.7
General expense	2,559 67	.27	7.6
Taxes and insurance	3,683 84	.39	11.0
Bullion expenses	569 39	.06	1.7
Total expense	$33,502 07	3.55	100.

Gross returns per cubic yard, 14.99 cents; net returns per cubic yard, 11.44 cents.

This dredge is working in fairly compact gravel with a heavy overburden.

7½-CUBIC-FOOT CLOSE-CONNECTED-BUCKET ELEVATOR DREDGE.

California type.

Summary of operations for two years eleven months and twelve days.

Actual dredging time, 13,464 hours 55 minutes; acres dredged, 54.649; cubic yards excavated, 3,458,229; average depth of ground, 27.9 feet.

	Total Costs.	Cost per Cubic Yard in Cents.
Operating, labor and material	$21,899 24	_____
Electricity	14,045 34	_____
Water	_____	_____
Repairs	51,859 55	_____
General expense	8,333 30	_____
Taxes and insurance	8,542 77	_____
Bullion expenses	941 19	_____
Total expense	$108,621 39	4.42

Gross returns per cubic yard, 9.30 cents; net returns per cubic yard, 4.88 cents.

These costs include depreciation and all fixed charges. The dredge is working in average bench gravel.

7½-CUBIC-FOOT DREDGE.

Statement showing the result of operation for the third year in operation. California type.

Operating expense.

Month Ending	Actual Dredging Time	Cubic Yards Excavated	Average Depth, Feet	Labor and Material	Electric Power	Repairs	General Expense	Bullion Expense	Taxes and Insurance	Total Expense	Total Cost Cubic Yard	Gross Returns	Gross Returns Cu. Yd.
	Hrs. Min.												
January 31	499 00	105,667	28.0	$1,185 04	$590 40	$2,165 41	$315 00	$33 78		$4,289 63	4.06	$9,580 74	9.07
February 29	536 25	96,213	28.5	1,358 31	633 60	1,703 14	283 12	31 64		4,009 81	4.17	7,525 62	7.82
March 31	656 55	119,354	25.8	1,429 35	637 20	6,526 69	304 57	38 36		8,936 17	7.49	9,326 78	7.81
April 30	540 25	120,191	26.6	1,155 21	615 60	1,823 01	266 77	61 07		3,921 66	3.26	13,086 24	10.89
May 31	627 30	122,875	27.0	1,142 70	684 00	1,758 70	291 47	57 83		3,934 70	3.20	15,943 17	12.98
June 30	626 20	123,665	30.0	1,090 06	757 80	4,462 49	333 03	41 98		6,685 36	5.40	9,497 11	7.68
July 31	645 40	107,073	32.0	973 89	691 20	3,900 01	238 72	36 21		5,840 03	5.45	9,029 62	8.43
August 31	537 35	92,948	29.0	1,079 06	588 60	1,891 77	265 92	25 89		3,851 24	4.14	6,814 93	7.33
September 30	351 10	42,542	27.0	817 11	415 80	2,020 03	273 00	23 12		3,549 06	8.34	4,298 75	11.28
October 31	534 20	65,821	27.5	1,074 66	606 60	1,630 63	260 96	27 48		3,600 33	5.47	6,739 40	10.24
November 30	533 35	94,439	30.5	1,099 71	725 40	1,515 20	404 25	29 80		3,774 36	4.00	7,645 83	8.10
December 31	611 50	103,358	28.5	87 74		3,791 84	164 64	41 39		4,085 61	3.95	10,470 85	10.13
									$4,338 00	4,338 00			
Totals	6,700 45	1,194,146		$12,492 84	$6,946 20	$33,188 92	$3,401 45	$418 55	$4,338 00	$60,815 96		$110,459 04	
Averages	Daily 18	Daily 3,264	Actual 27.5										
Cost per cubic yard, cents				1.05	.58	2.78	.28	.04	.37		5.10		9.25
Per cent of total cost				20.6	11.4	54.5	5.6	0.7	7.2	100.0			

These costs include depreciation and all fixed charges. Formation average bench gravel.

7½-CUBIC-FOOT CLOSE-CONNECTED-BUCKET ELEVATOR DREDGE.

Two and one half years in commission. California type.

Summary of working costs for the last twelve months in operation.

Actual dredging time, 6,900 hours 50 minutes; cubic yards excavated, 1,281,351; average depth of ground, 67.8 feet.

	Total Costs.	Cost per Cubic Yard in Cents.
Operating, labor	$13,906 70	
Electricity	12,562 57	
Water		
Repairs, labor and material	25,730 67	
General expense	5,856 56	
Total expense	$58,056 50	4.53

These costs include depreciation and all fixed charges. This dredge is working in medium light gravel overlain by hydraulic tailing.

7½-CUBIC-FOOT CLOSE-CONNECTED-BUCKET ELEVATOR DREDGE.

Two and one half years in commission. California type.

Summary of working costs for the last twelve months in operation.

Actual dredging time, 6,402 hours 35 minutes; cubic yards excavated, 1,369,844; average depth of ground, 70.2 feet.

	Total Costs.	Cost per Cubic Yard in Cents.
Operating, labor	$13,375 88	
Electricity	10,596 44	
Water		
Repairs, labor and material	26,522 49	
General expense	6,357 08	
Total expense	$56,851 89	4.15

These costs include depreciation and all fixed charges. This dredge is working in medium light gravel overlain by hydraulic tailing.

8-CUBIC-FOOT CLOSE-CONNECTED-BUCKET ELEVATOR DREDGE.

Four months and eight days in commission. California type.

Working costs.

Days operating, 130; actual dredging time, 2,369 hours 5 minutes; cubic yards excavated, 626,264; average depth of ground, 24 feet; acres dredged, 16.38.

	Total Costs.	Cost per Cubic Yard in Cents.
Operating, labor and material	$6,846 02	
Electricity	3,326 00	
Water		
Repairs	2,524 44	
General expense	1,241 38	
Taxes and insurance	1,237 18	
Bullion expenses	280 84	
Total expense	$15,455 86	2.47

Gross returns per cubic yard, 11.72 cents; net returns per cubic yard, 9.25 cents.

These costs include depreciation and all fixed charges. The dredge is digging average bench gravel.

8-CUBIC-FOOT CLOSE-CONNECTED-BUCKET ELEVATOR DREDGE.

Six months in commission. California type.

Summary of working costs.

Days operating, 171; actual dredging time, 3,162 hours 5 minutes; cubic yards excavated, 583,927; acres dredged, 8½; average depth of ground, 42.5 feet.

	Total Costs.	Cost per Cubic Yard in Cents.
Operating, labor and material	$9,879 13	--------
Electricity	3,470 40	--------
Repairs	6,343 22	--------
General expense	1,665 14	--------
Taxes and insurance	1,383 91	--------
Bullion expenses	302 58	--------
Total expense	$23,044 38	3.95

Gross returns per cubic yard, 12.67 cents; net returns per cubic yard, 8.72 cents.

These costs include depreciation and all fixed charges. The dredge is operating in light gravel and against a bank about 10 feet above the water line.

9-CUBIC-FOOT CLOSE-CONNECTED-BUCKET ELEVATOR DREDGE.

California type.

Working costs for the first five months in operation.

Cubic yards excavated ---- 580,310
Cubic yards per month ---- 116,062
Average depth of ground, feet ---- 51
Total cost ---- $29,009.18
Cost per cubic yards in cents ---- 4.98

These costs include all fixed charges. This dredge is digging under most difficult conditions in cemented gravel and against a bank 20 feet above the water level.

13½-CUBIC-FOOT CLOSE-CONNECTED-BUCKET ELEVATOR DREDGE.

Nearly eight months in commission. California type.

Working costs.

Days operating, 235; actual dredging time, 4,478 hours 20 minutes; cubic yards excavated, 1,830,201; acres dredged, 60.02; average depth of ground, 19 feet.

	Total Costs.	Cost per Cubic Yard in Cents.
Operating, labor and material	$18,690 91	--------
Electricity	8,618 70	--------
Water		--------
Repairs	10,246 87	--------
General expense	2,114 02	--------
Taxes and insurance	1,759 25	--------
Bullion expenses	736 09	--------
Total expense	$42,165 84	2.3

These costs include all operating costs, depreciation, insurances, tax, and all fixed charges. This machine is digging in fine gravel wash.

Capacity of Buckets. Cubic Feet.	Time in Commission for Figures Given. Year.	Month.	*Working Period for Figures Given. Year.	Month.	Time Dredging. Total. Hours.	Min.	Daily. Hours.	Min.	Area Dredged. Acres.	Depth of Gr... in Feet. Average.	Min.
5	2	5	1	---	6,790	35	18	36	23.87	29.9	---
5	4	*6	1	---	6,507	20	---	---	13.96	36.0	
5	6	5	1	---	6,513	00	17	48	11.86	30.9	2
5	6	---	3	---	19,463	10	---	---	43.94	29.5	
5	7	6	1	---	6,015	20	16	34	16.75	25.4	---
5	3	6	1	---	6,798	45	18	35	27.91	25.5	
7	---	10	---	10	5,309	50	14	06	12.05	32.2	---
7	1	3	1	---	6,313	35	17	18	25.05	27.6	---
7	3	9	1	---	6,390	20	17	28	24.23	26.5	
7½	2	9	1	---	6,700	45	18	18	26.92	27.5	2
7½	---	10	---	10	5,582	15	19	52	20.42	28.7	2
7½	2	8	2	8	13,464	55	18	20	54.64	27.9	---
7½	2	6	1	---	6,402	35	---	---	---	70.2	---
7½	3	9	1	---	6,651	55	18	19	28.95	26.9	---
7½	1	10	1	---	7,048	10	19	25	24.75	30.22	2
8	1	4	1	---	6,849	05	18	52	51.88	23.44	1
8	1	6	1	---	6,886	40	18	58	25.19	39.47	3
8½	5	---	1	---	7,037	38	19	36	43.17	24.73	1
9	4	---	1	---	5,763	45	16	03	15.80	52.51	4
9	1	9	1	---	6,594	25	18	22	18.93	51.25	4
13	4	1	---	1	445	15	---	---	---	25.2	---
13	4	1	---	1	467	21	---	---	---	23.1	
13	4	1	---	1	469	08	---	---	---	26.1	---
13	4	1	---	1	537	56	---	---	---	20.6	
13	4	1	---	1	584	32	---	---	---	19.0	---
13	4	1	---	1	575	06	---	---	---	20.5	
13	4	1	---	1	616	40	---	---	---	17.5	---
13	4	1	---	1	640	21	---	---	---	19.9	---
13	4	1	---	1	598	38	---	---	---	24.1	---
13	4	1	---	1	615	47	---	---	---	23.2	---
13	4	1	---	1	524	02	---	---	---	24.2	---
13	4	1	---	1	584	34	---	---	---	24.9	
13	4	1	1	---	6,659	20	18	33	77.43	21.87	---
13½	1	7	---	1	645	45	---	---	---	16.1	---
13½	1	7	---	1	575	50	---	---	---	20.0	---
13½	1	7	---	1	603	00	---	---	---	18.6	---
13½	1	7	---	1	536	53	---	---	---	19.3	---
13½	1	7	---	1	518	58	---	---	---	19.6	---
13½	1	7	---	1	609	05	---	---	---	19.4	---
13½	1	7	---	1	579	45	---	---	---	19.7	---
13½	1	7	---	1	647	50	---	---	---	22.1	---
13½	1	7	---	1	587	40	---	---	---	18.7	---
13½	1	7	---	1	595	35	---	---	---	18.2	---
13½	1	7	---	1	559	55	---	---	---	18.3	---
13½	1	7	---	1	509	00	---	---	---	19.3	---
13½	1	7	1	---	6,969	13	19	12	99.05	19.07	---

* The last year or months in operation at end of time in commission for figures given.　　　† Labor o

TABLE SHOWING WORKING COSTS OF CLOSE-CONNECTED-BUC...

| | | Ground Worked. | | Total Cost Per Cubic Yard in Cents. | RUNNING EXPENSES. | | | | | |
| ound | | Total. | Daily. | | Labor and Material. | | Material Only. | | Electric Pow... | |
	Max.	Cubic Yards.	Cubic Yards.		Cost Per Cubic Yard in Cents.	Per Cent of Total Cost.	Cost Per Cubic Yard in Cents.	Per Cent of Total Cost.	Cost Per Cubic Yard in Cents.	Pe. Tot.
		1,148,802	3,148	3.64	.82	22.7			.49	
		812,355		6.65	2.30	34.56			1.08	
3	33	592,078	1,618	6.50	1.75	27.0			.89	
		2,089,163		5.59						
		679,572	1,872	5.65	1.48	26.2	.22	3.8	.73	
		1,148,480	3,138	3.80	.88	23.1			.52	
		645,095	2,122	5.41	1.87	34.5			1.12	
		1,114,605	3,054	4.07	1.21	29.7			.62	
		1,033,694	2,824	5.09	1.08	21.2			.64	
6	32	1,194,146	3,264	5.10	1.05	20.6			.58	
5	33	944,879	3,363	3.55	.95	26.7			.58	
		2,458,229		4.42						
		1,369,844		4.16	.99				.77	
8	34	1,257,055	3,463	4.78	†1.28	26.67	.06	1.4	.74	
9	34	1,177,772	3,244	4.03	1.17	28.9	.05	1.4	.68	
5	51	1,962,448	5,406	3.11	.98	31.2	.13	4.2	.65	
7	30	1,604,369	4,419	4.05	1.01	24.8	.15	3.7	.74	
7	61	1,722,281	4,791	4.33	.79	18.3	.09	2.3	.93	
7	70	1,339,464	3,710	5.51	.91	16.6	.19	3.4	.93	
		1,565,598	4,361	5.86	.88	15.0	.15	2.6	1.05	
		170,830		4.09						
		177,974		3.43						
		204,500		3.02						
		201,831		2.04						
		231,375		2.25						
		236,533		2.04						
		274,318		3.13						
		295,051		1.84						
		221,014		3.16						
		195,403		4.84						
		231,885		3.25						
		292,057		1.98						
		2,732,771	7,612	2.77	.62	22.3	.06	2.5	.44	
		309,741		2.07						
		285,329		2.20						
		308,494		2.64						
		231,754		2.81						
		233,010		3.31						
		256,758		2.49						
		239,758		1.97						
		257,995		2.56						
		218,189		2.45						
		255,826		2.55						
		230,636		1.60						
		220,764		2.33						
		3,048,254	8,397	2.41	.60	24.7	.15	6.1	.54	

nly.

			REPAIRS.				General Expenses.		Ta
			Labor and Material.		Material Only.				
ver.	Water.								
r Cent of al Cost.	Cost Per Cubic Yard in Cents.	Per Cent of Total Cost.	Cost Per Cubic Yard in Cents.	Per Cent of Total Cost.	Cost Per Cubic Yard in Cents.	Per Cent of Total Cost.	Cost per Cubic Yard in Cents.	Per Cent of Total Cost.	C Y
13.4	.03	0.7	1.89	52.0			.25	6.9	
16.23			2.92	43.89					
13.6			2.86	43.9			.57	8.9	
12.8			.93	16.4	1.22	21.6	.69	12.3	
13.8	.05	1.4	1.77	46.4			.25	6.7	
20.7			1.77	32.8			.51	9.4	
15.3	.03	0.7	1.81	44.5			.29	7.0	
12.6	.14	2.9	2.69	52.8			.34	6.6	
11.4			2.78	54.5			.28	5.6	
16.3			1.30	36.7			.27	7.6	
			1.95				.45		
15.4			†.28	5.8	1.82	37.9	.37	7.8	
16.8			.33	8.0	1.14	28.2	.40	10.0	
20.9	.02	0.9	.15	4.8	.83	26.7	.22	7.0	
18.3	.11	2.8	.26	6.5	1.34	33.1	.27	6.6	
21.6	.20	4.6	.47	10.8	1.00	22.9	.56	13.0	
16.9	.15	3.2	.89	16.2	1.37	24.8	.69	12.6	
17.9	.09	1.7	.57	9.7	2.21	37.7	.58	9.9	
15.8	.12	4.1	.27	9.8	.75	27.1	.33	11.9	
21.9	.08	3.6	.10	4.4	.70	28.9	5.14	5.8	

xes and Insurance.		Smelting and Express Charges.		REMARKS.
ost per Cubic Yard in Cents.	Per Cent of Total Cost.	Cost per Cubic Yard in Cents.	Per Cent of Total Cost.	
.16	4.3			Operating in very favorable ground.
.35	5.31			Operating in average fair ground.
.39	6.0	.04	.06	Operating in average fair ground.
				Operating under favorable conditions.
.35	6.3	.03	.6	Operating in fair average ground; dredge partly stopped for repairs.
.33	8.6			Digging under favorable conditions.
.14	2.6			Digging in very difficult ground.
.11	2.8			Compact gravel.
.20	3.9			Difficult ground; heavy replacement charges.
.37	7.2	.04	0.7	Average fair ground; dredge partly shut down in February, August, September, October, and November.
.39	11.0	.06	1.7	Average fair ground; dredge partly shut down in August, September, and October.
				Average digging ground.
				Average digging ground.
.19	4.1	.04	.9	Average digging ground.
.21	5.2	.05	1.5	Average digging ground.
.10	3.2	.03	1.1	Fairly tight; free washing ground.
.12	3.1	.05	1.1	Fairly tight; free washing ground.
.25	5.6	.04	.9	Fairly tight; free washing ground.
.31	5.6	.04	.7	Light and partly cemented gravel.
.26	4.6	.07	.9	Light and partly cemented gravel.
.15	5.5	.03	1.0	Fairly loose gravel.
.06	2.7	.04	1.9	Fairly loose gravel.

STATEMENT SHOWING LOST TIME.

5-cubic-foot close-connected-bucket elevator dredge. Five and one half years in operation.

Lost time for the last twelve months in operation.

	Total Time Lost.
Bucket line, ladder and tumblers	198 hours 35 minutes
Screens	76 hours 55 minutes
Stacker	32 hours 10 minutes
Pumps	5 hours 50 minutes
Winches	26 hours 45 minutes
Lines	89 hours 40 minutes
Power	283 hours 10 minutes
Stepping ahead	298 hours
Clean-ups	56 hours
Other causes	736 hours 10 minutes
Total time lost	1,803 hours 15 minutes
Total time running	3,980 hours 45 minutes
Total time possible	8,784 hours

STATEMENT SHOWING LOST TIME.

3½-cubic-foot close-connected-bucket elevator dredge. Six and one half years in operation.

Total time lost for the last twelve months in operation.

	Total Time Lost.	Per Cent of Total.
Bucket line	458 hours 25 minutes	5.22
Screens	40 hours 30 minutes	.46
Stacker	29 hours	.33
Pumps	39 hours 40 minutes	.45
Winches	57 hours 30 minutes	.66
Lines	14 hours 40 minutes	.16
Power	396 hours 20 minutes	4.51
Stepping ahead	84 hours 55 minutes	.97
Clean-ups	50 hours 45 minutes	.58
Other causes	406 hours 35 minutes	4.63
Total time lost	1,578 hours 20 minutes	
Total time running	7,205 hours 40 minutes	
Total time possible	8,784 hours	

STATEMENT SHOWING LOST TIME.

5-cubic-foot close-connected-bucket elevator dredge. Four and one half years in commission.

Lost time for the last twelve months in operation.

	Total Time Lost.	Per Cent of Total Lost Time.
Bucket line	161 hours 5 minutes	1.834
Tumblers	212 hours 35 minutes	2.420
Screens	81 hours 45 minutes	.931
Ladder	502 hours 10 minutes	5.717
Pumps	83 hours 5 minutes	.946
Winches	95 hours 10 minutes	1.083
Lines	116 hours 50 minutes	1.330
Spuds	28 hours	.319
Motors	18 hours 55 minutes	.215
Power	420 hours 30 minutes	4.787
Stacker	239 hours 5 minutes	2.722
Stepping ahead	164 hours 10 minutes	1.869
Clean-ups	64 hours 45 minutes	.737
Stumps	55 minutes	.010
Other causes	87 hours 40 minutes	.998
Total time lost	2,276 hours 40 minutes	25.918
Total time running	6,507 hours 20 minutes	74.082
Total time possible	8,784 hours	100.

8-CUBIC-FOOT CLOSE-CONNECTED-BUCKET ELEVATOR DREDGE.

Average weekly record of the Natoma No. 2 during the first seven months in operation.

NATOMAS CONSOLIDATED OF CALIFORNIA—NATOMA DIVISION.

Date.	Actual Dredging Hrs. Min.	Bucket Line	Upper Tumbler	Lower Tumbler	Screens	Conveyor	Water Pumps	Sand Pumps	Side-line Winch	Ladder Winch	Side Lines	Spuds	Motors	Oiling	Power Off	Main Gearing	Stepping Ahead	Clean-ups	Transformers, etc.	Depth, Feet	Lost Time Hrs. Min.
February 25	12-40	3-15			1-55	0-45					0-10			0-15				4-10	1-15	25.1	11-20
February 26	21-35					0-35	0-25							0-15	1-0		0-25			26.1	2-25
February 27	22-10				0-05	0-25	0-20							0-15	0-15		0-25			26.9	1-50
February 28	21-10				1-20						0-45			0-15			0-10			26.7	3-50
March 1	20-20	0-25			0-20	1-30					0-15			0-15	0-10		0-40		0-25	28.2	3-40
March 2	20-35	0-40				0-25					0-15			0-15			0-25			27.0	3-25
March 3	22-25	0-40					1-05										0-25		0-15	27.5	1-35
Totals	140-55	5-0			3-40	3-40	1-50				1-25			1-30	1-25		2-30	4-10	1-55		27-05
Daily average	20-08																				
Average depth																				26.8	

CAUSE OF LOST TIME.

Ground dredged for week (7 days), 47,635 cubic yards; ground dredged per day (24 hours), 6,719.3 cubic yards; ground dredged per hour (7 days of 24 hours each), 279.9 cubic yards; ground dredged per hour of actual dredging time, 333.8 cubic yards; power used, 37,920 kilowatt hours.

13½-CUBIC-FOOT CLOSE-CONNECTED-BUCKET ELEVATOR DREDGE.

Table showing an average weekly record of the Natoma No. 1 during the first twelve months in operation.

NATOMAS CONSOLIDATED OF CALIFORNIA—NATOMA DIVISION.

Date	Actual Dredging Hrs. Min.		Cause of Lost Time																			Depth, Feet	Lost Time Hrs. Min.	
	Hrs.	Min.	Bucket Line	Upper Tumbler	Lower Tumbler	Screens	Conveyor	Water Pumps	Sand Pumps	Side-line Winch	Ladder Winch	Side Lines	Spuds	Motors	Oiling	Power Off	Main Gearing	Stepping Ahead	Clean-ups	Power Cable	Hopper		Hrs.	Min.
March 24	13	50	1-0		0-30	0-30	1-0			0-30		0-20						0-10	6-10			19.3	10	10
March 25	20	35				1-15	0-10				0-10	1-10			0-25			0-25	0-05	1-10		19.0	3	25
March 26	10	55			8-0			3-0			0-05				0-15			0-30				19.6	13	05
March 27	21	45				0-05	0-15				0-10				0-30			0-50			0-15	18.6	2	15
March 28	22	40										0-25			0-30			0-10				19.0	1	20
March 29	22	35										0-10			0-35		0-15	0-10				18.3	1	25
March 30	22	45										0-20			0-35			0-20				18.7	1	15
Totals	135	05	1-0		8-30	2-05	1-35	3-0		0-30	0-25	2-25			2-50		0-15	2-40	6-15	1-10	0-15		32	55
Daily average	19	18																						
Average depth																						18.93		

Ground dredged for week (7 days), 72.151 cubic yards; ground dredged per day (24 hours), 10,307.3 cubic yards; ground dredged per hour (7 days of 24 hours each), 429.4 cubic yards; ground dredged per hour of actual dredging time, 534.1 cubic yards. *Remarks*—New lower tumbler 3-26-09. Power used, 42,560 kilowatt hours.

V.   CALIFORNIA DREDGING DISTRICTS.

1.   BUTTE COUNTY.

Butte County, comprising an area of 1,720 square miles, or 1,100,800 acres, with a population of about 30,000 inhabitants, is situated about 150 miles from San Francisco.   It is bounded on the north and east by Tehama and Plumas counties, on the south by Yuba and Sutter, and on the west by Colusa and Glenn counties.

Oroville, the county seat, with a population of 6,000 is located on the Feather River.   It is on the main line of the Western Pacific and is

No. 81.   View of Feather River, near Oroville.

reached by branch lines of the Southern Pacific and Northern Electric railroads.

There are two rivers in the county, the Sacramento and Feather.   The Sacramento River forms the western boundary, and is navigable 100 miles north of Butte County, and south, its entire length to San Francisco Bay.   The Feather River drains an area of about 3,640 square miles, and flows through the center of the county, supplying water to a number of irrigation systems, the most important being the Butte County Canal Company's and the Feather River Canal Company's ditches.

In the northern and eastern parts of the county, which are mountainous and covered with timber, lumbering, stock raising and mining are

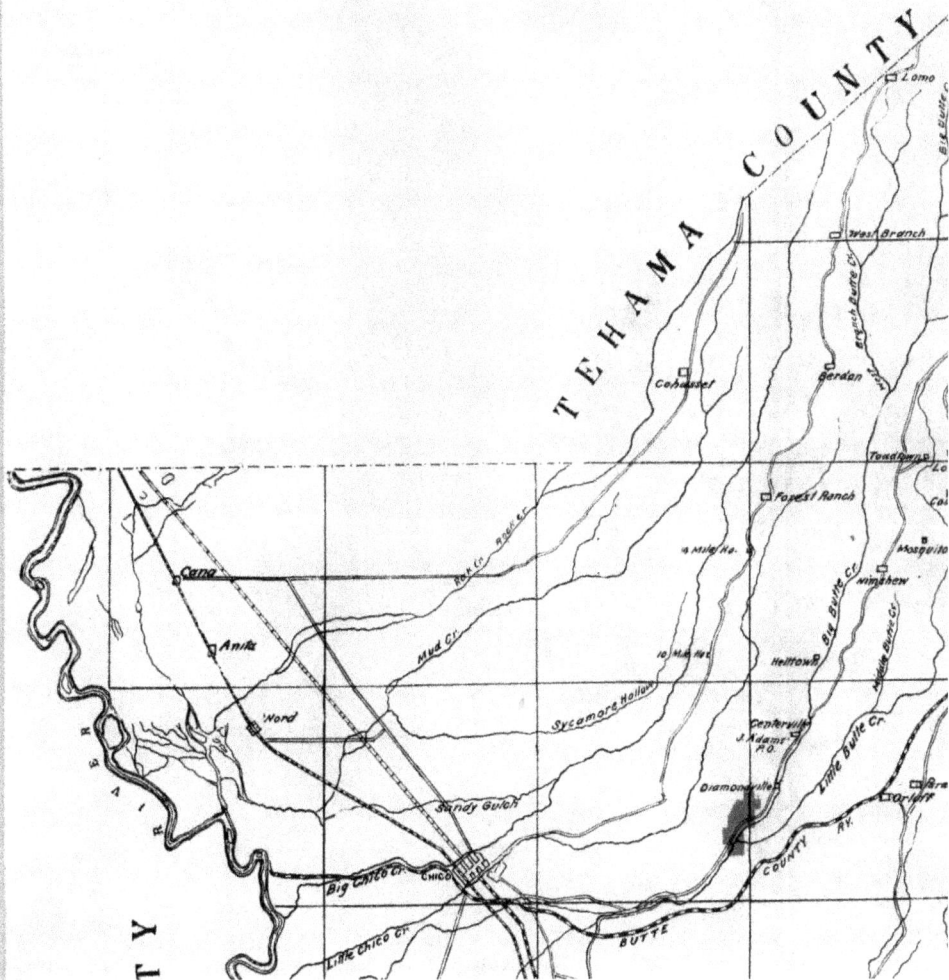

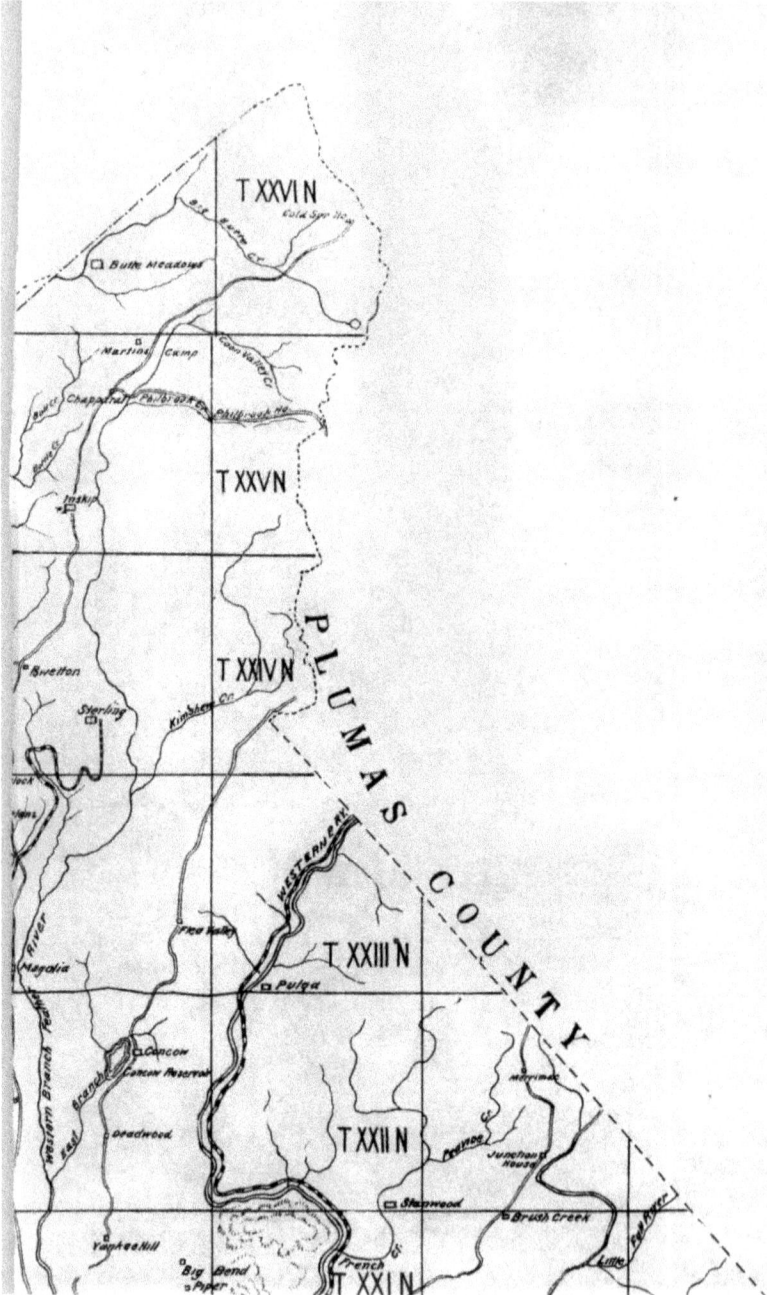

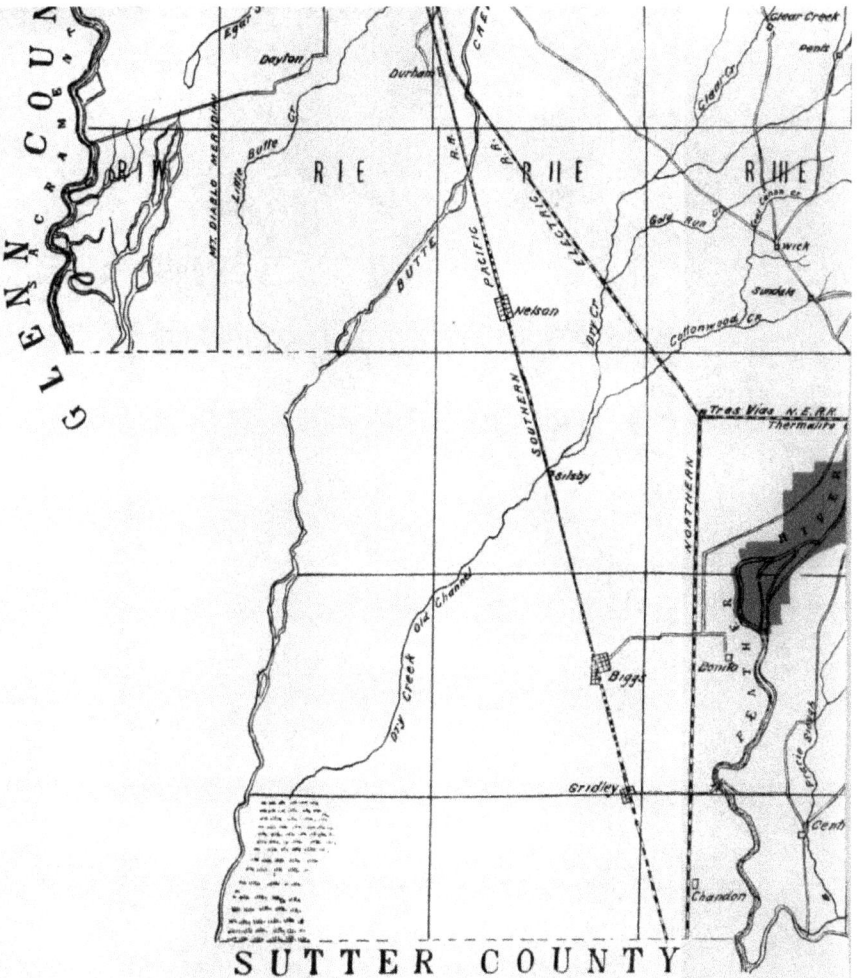

MAP OF BUTTE COUNTY, CALIFORNIA, SHOWING

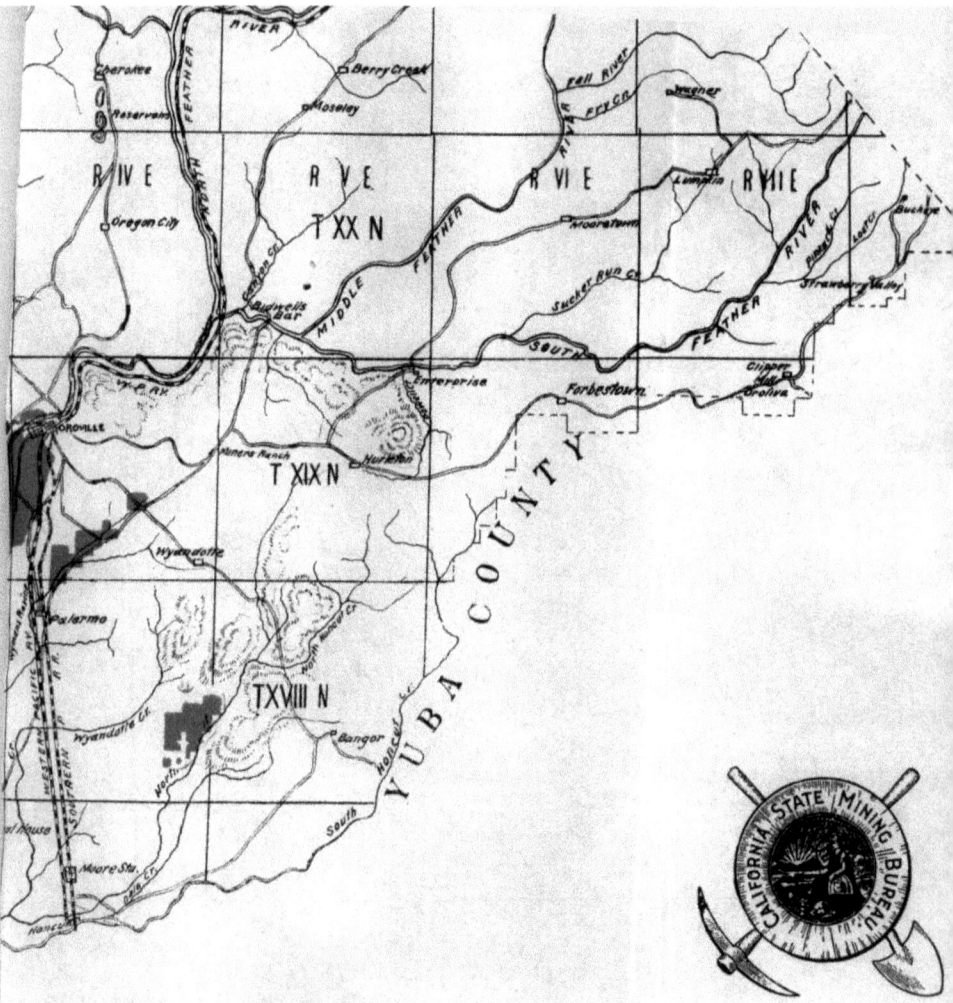

LOCATION OF DREDGING LAND.

carried on, while in the valley and foothill regions, gold dredging is the leading industry, fruit growing and agriculture being also extensively carried on.

Butte County is well supplied with transportation facilities. The Southern Pacific, Western Pacific, and Northern Electric railroads traverse the county, and river steamers run regular schedules on the Sacramento River.

There are 1,600 miles of public roads in the county, 152 miles of electric power lines, four electric power plants, 442 miles of irrigating ditches, and 14,000 acres of land under irrigation.

The following table shows the mineral production of Butte County from 1900 to 1908:

	1900.	1901.	1902.	1903.	1904.
Brick	------	$7,200	$5,000	$7,200	$4,020
Gold	$485,589	864,978	916,782	1,571,507	1,932,552
Lime	600	1,500	750	250	------
Limestone	------	------	------	250	------
Macadam	------	------	------	------	------
Mineral water	1,515	1,455	1,500	1,550	1,512
Platinum	------	------	------	210	1,000
Silver	13,082	4,634	2,219	358	2,302
Unapportioned	------	------	------	------	------
Totals	$500,786	$879,767	$926,251	$1,581,325	$1,941,386

	1905.	1906.	1907.	1908.	
Brick	$3,200	$1,300	------	------	------
Gold	2,607,500	3,016,747	$2,786,840	$3,139,398	------
Lime	------	------	------	------	------
Limestone	------	------	------	------	------
Macadam	------	------	------	7,916	------
Mineral water	1,500	1,950	2,140	2,450	------
Platinum	1,770	475	------	------	------
Silver	7,134	10,853	8,967	12,708	------
Unapportioned	------	------	------	------	$105,870
Totals	$2,621,104	$3,031,325	$2,797,947	$3,162,472	$17,548,233

Nearly all dredge mining in Butte County is carried on in the Oroville or Feather River district on land adjacent to the Feather River. The district, comprising a dredgeable area of about 6,450 acres, nearly two thirds of which has been already dredged, includes in its boundaries part of Oroville and extends six miles south of the town, ranging from one to several miles in width. Some minor dredging areas in Butte County are located near Oroville on Wyman's Ravine and Honcut Creek, tributaries of the Feather River, and on Butte Creek near Diamondville. As dredging in California began in the Oroville district, it may be of interest to give in detail some of the early history of the industry.

W. P. Hammon and Warren Treat were among the first to attempt working the gravels for gold on a large scale or by different methods than the sluice box or rocker. Treat, in the summer of 1895, sunk a pit about 100 feet square to bedrock, using a centrifugal pump to handle the water, and in spite of the heavy costs caused by the rehandling of the gravel several times by manual labor and for pumping, made a profit. Hammon working another pit found the heavy flow of water on approaching bedrock prevented profitable working of the lower gravel, but from the values recovered, realized the great importance of the gravel fields if economic methods could be utilized, and had for some time thought of the then little known process of gold dredging in connection with same. He accordingly secured favorable options on a large area of ground in the vicinity of Oroville. Thomas Couch, a Montana mining man, was interested in the venture by Frank T. Southerland, and agreed to have a

No. 32. Treat's old workings at Oroville, showing bedrock covered by water, making dredge-mining the only possible system for working the gravels.

thorough test made of the gravel and to finance the propositions if the results proved satisfactory.

The manner of prospecting at that time was crude as compared to present methods, a couple of Chinamen with picks, shovels, pans, and rocker, comprised the outfit. A shaft was sunk until water-level was reached and the gravel from same put through a rocker, the tailing from the rocker being carefully panned. The results obtained from a number of shafts sunk at various parts of the field were so satisfactory that Captain Couch said that if the gravel contained gold below the

No. 83. Remnants of the first dredge at Oroville, the Feather River or Couch No. 1, Oroville District, 1909.

water-level in proportion to that found above, he would feel justified in ordering the construction of a dredge. In order to determine the value of the gravel below water-level, permission was secured to use the pumping plant of Treat, the pit being unwatered and the sides sampled. The results being satisfactory, the contract was made for the first dredge, the Couch No. 1, which after numerous changes and repairs proved a success. Soon after this, the coöperation of the Lewishon Brothers of New York was secured, and the Feather River Exploration Company organized; practically all the stock of this company was held by the Lewishons, Couch, Hammon, and Southerland.

While much credit is due to Captain Couch, who has been thought by many to be the principal figure in the early history of dredging in California, it must not be overlooked that the whole scheme originated with W. P. Hammon, and it was his enterprise and faith in the project that encouraged the others and kept the work going in spite of repeated setbacks, sufficiently discouraging to overwhelm men of great determination. John J. Hamlyn, the first secretary of the Feather River Exploration Company, was another pioneer in this field, and took an active part in its development.

There is a great deal to be said regarding the development of the dredging industry in the Oroville district, and many to whom credit should be given for their part in the development of the California dredge, but time and space does not permit. It is a well known fact that dredging in the Oroville district has been a great financial success, and for this reason, the following table showing the production of gold as nearly as can be ascertained during the first ten years of dredging operations may be of interest:

Year.	Amount.	Decrease.	Year.	Amount.	Decrease.
1898	$18,847		1905	$2,261,887	
1899	132,412		1906	2,768,782	
1900	154,065		1907	2,697,092	$71,690
1901	396,919		1908	3,043,051	
1902	614,380				
1903	1,329,998		Total	$15,049,940	
1904	1,632,507				

The decrease in 1907 as compared with 1906 was partly due to two dredges being completely wrecked and others more or less damaged by

No. 84. The old Oroville and California Company's Marion dipper dredges, wrecked by floods in 1907. Oroville District.

floods during the spring of 1907, so that few of the dredges operated continuously during the entire year.

During 1908, there were 35 dredges and 12 dredging companies operating in the Oroville district. These were as follows: Indiana Gold Dredging Company, 1; Butte Dredging Company, 1; El Oro Dredging Company, 2; Ophir Gold Dredging Company, 2; Gold Run Dredging Company, 1; Viloro Syndicate, Limited, 1; Oro Water, Light and Power Company, 6; Pennsylvania Dredging Company, 1; Oroville Dredging, Limited, 9; Feather River Development Company, 6 dredges.

DREDGING IN THE OROVILLE DISTRICT.

At the beginning of the year 1909 there were thirteen companies in the field, controlling about 6,450 acres and operating 30 dredges.

During the year one dredge was wrecked by floods, one reconstructed, one destroyed by fire, one dismantled, and one put out of commission. The following table shows the numerical strength of the dredges in the Oroville district during the first and last half of 1909:

	Operat-ing.	Recon-structing.	Wrecked.	Dis-mantled or Closed Down.
First half of 1909	30	1	1	1
Last half of 1909	28	--	1	1

The dredges wrecked were the Indiana No. 3, which was partly destroyed by the floods of January 15th and later reconstructed and put

No. 85. Old Risdon dredge in operation. The Marigold No. 1, now dismantled.

in commission in July, and the Viloro No. 1, which was destroyed by fire September 2d. The dredges dismantled or put out of commission were the Leggett No. 3, which was dismantled during the month of June, and the Continental, which was put out of commission during the month of November, having worked out the property. A number of the operating dredges were laid up for repairs during the year; among these were the Empire, which turned over in the pond owing to a leak in the hull, and, in consequence, was idle for the better part of two months. The Pennsylvania and Butte dredges were idle for some time, owing to repairs to spuds and stackers, etc.

The heavy storms during the early part of the year delayed operations on some of the dredges, while minor repairs caused stoppages on others, so that few of the dredges operated continuously during the entire year.

The following table gives the names of the companies and dredges operating in Butte County in 1909, as well as the type of the different

No. 86.   Remnants of one of the early dredges in the Oroville District.   Old Risdon type.

No. 87.   The Continental Dredge in 1909, after ten years in operation, four months before being dismantled.   Oroville District.

dredges, the capacity of the buckets and the time the various dredges have been in operation up to January 1, 1910:

Company.	Dredge.	Type of Dredge.	Date Began Operations.	Date Stopped Operations.	Time in Operation. yr. mo.	Capacity of Buckets, Cubic Feet.
Oroville District.				1909.		
Indiana Gold Drdg. and Min. Co.	Indiana No. 3	Close C.	July, '09		6	4
Butte Dredging Co.	Butte No. 1	Close C.	Nov. 26,'02		7 1	3½
El Oro Dredging Co.	El Oro No. 1	Close C.	Dec. 26,'03		6 11	5
	El Oro No. 2	Close C.	Jan. 22,'08		11	5
Ophir Gold Dredging Co.	Ophir	Close C.	Aug. 4,'06		3 5	5
	Nevada	Close C.	May 15,'04		5 7½	4
Gardella	Gardella	Open C.	May 1,'02	Aug.	7 4	5
Gold Run Dredging Co.	Baggett No. 1	Open C.	May 26,'06		3 9	7
Viloro Syndicate, Ltd.	Viloro No. 1	Close C.	Oct. 30,'04	Sept. 2	4 10	5
Oro Water, Light and Power Co.	Lava Bed No. 2	Close C.	July, '03		6 6	5
	Lava Bed No. 3	Close C.	Dec. 4,'04		5 1	5
	Empire	Close C.	April 22,'06		3 8	5
	Victor	Close C.	Sept. 4,'07		2 4	5
	Hunter	Close C.	Aug. 13,'07		2 4½	5
Pennsylvania Dredging Co.	Pennsylvania	Close C.	Nov., '02		7 2	6
Oroville Dredging, Ltd.	California No. 2	Close C.	Jan. 1,'03		6	5
	California No. 3	Close C.	Oct., '04		5 2½	7
	Explorat'n No.1	Open C.	April, '02		7 8½	3
	Explorat'n No.2	Close C.	Feb. 17,'05		4 10½	5
	Explorat'n No.3	Close C.	Oct. 20,'06		3 2⅝	7
	Boston No. 4	Close C.	May 2,'06		3 7½	7
	Continental	Close C.	June, 1899	Nov.	10 4	4
Natomas Cons. of California	Feather No.1	Close C.	July, '02	Dec. 31	7 6	5
(Feather River Division)	(Cherokee)					
	Feather No. 2	Close C.	Dec. 22,'06		3	7½
	Feather No. 3	Close C.	Mar. 26,'08		1 9	7½
	Feather No. 4	Open C.	Dec. 10,'02	Dec. 31	7	3½
Pacific Gold Dredging Co.	Pacific No. 1	Close C.	May 1,'06		3	7½
	Pacific No. 2	Close C.	——, '02		7 6	4
	Pacific No. 3	Close C.	April, '04		5 8	5
	Pacific No. 4	Close C.	Jan. 26,'08		1 11	7
Leggett Gold Dredging Co.	Leggett No. 3	Open C.	April, '04	June	5 2	5
Wyman's Ravine District.						
Leggett Mining Co.	Leggett No.4	Open C.	Sept., '09		4	5
L. & J. Gardella	Gardella	Open C.	Oct. 12,'07		2 2½	5¾
Honcut Creek District.						
Kentucky Ranch Gold Dredg. Co.	Kentucky	Open C.	April, '09		9	5
Butte Creek District.						
Butte Creek Cons. Dredging Co.	Butte Creek	Open C.	May 1,'09		8	11

NOTE.—Close C., Close connected; Open C., Open connected.

The production of gold from dredging operations in 1909 amounted to about $2,900,000 or $103,571 per dredge for the average number of working dredges.

The following table shows the number of dredges in Butte County in 1910:

District.	Dredges.			
	Operating.	Dismantled.	Constructing.	Total.
Oroville	25	--	1	26
Wyman's Ravine	2	--	--	2
Honcut Creek	2	--	--	2
Butte Creek	1	--	--	1
Totals	30	--	1	31

No. 88.   Hunter Dredge, 5-cubic-foot close-connected buckets, California type.   Oroville District.   Western Engineering and Construction Co. design.

In the Oroville district, on December 31, 1909, the Cherokee and Feather No. 4 dredges were permanently closed down. On the Cherokee property a 13½-cubic-foot dredge will be put in operation during 1910.

The Indiana Gold Dredging Company began operations in 1901 under the name of the Indiana Gold Dredging and Mining Company, and up to 1907 commissioned two dredges, the Indiana No. 1 and No. 2. In 1907 the company made some changes and reorganized under the laws of the State of Nevada, at the same time changing the name. The officers of the company are located at Oroville, California, and the names of the present officers are as follows: President, H. F. Perry; vice-president, J. F. Newson; secretary and manager, O. C. Perry; dredge superintendent, Luther Hadley.

The holdings of the company comprise 163 acres located in sections 13 and 14, township 19 north, range 3 east, and sections 18 and 19, township 19 north, range 4 east, all on the east side of the Feather River and about one half to one mile distant from the town of Oroville. The dredgeable area comprises about 140 acres, of which more than 100 acres have been dredged. More than two thirds of the dredging area had been mined by Chinese and whites by the processes in use before the advent of the dredge; the rest is overflow land located near the banks of the Feather River.

Owing to the location of the overflow land, the company has been burdened with unusual expenses and serious disasters to their dredging machines. On March 17, 1907, the Indiana No. 1 and No. 2 were wrecked by floods while working on low ground near the Feather River. The machinery of No. 1 was sold and put on a dredge operating at Snelling, Merced County, and from the wreck of No. 2, the Indiana No. 3 was constructed, which was put in operation in December of the same year. A little over a year later, this dredge was partly wrecked in the floods of January 15, 1909, and after being repaired and partly reconstructed it was recommissioned in July of the same year, and since then has been in active operation.

The operations of the Indiana Gold Dredging and Mining Company are of special interest and worthy of notice inasmuch as in the construction of the Indiana No. 1 dredge, the company, under the management of O. B. Perry, practically originated what is known to-day as the Bucyrus type of placer mining elevator dredge. To understand the difficulties confronting the designers of this dredge, one must know that the majority of the elevator dredges in use for placer mining in America, up to 1901, were confined to two distinct types of construction; the old style double-lift Bucyrus type of dredge, equipped with tail sluices and tail scow, and successfully

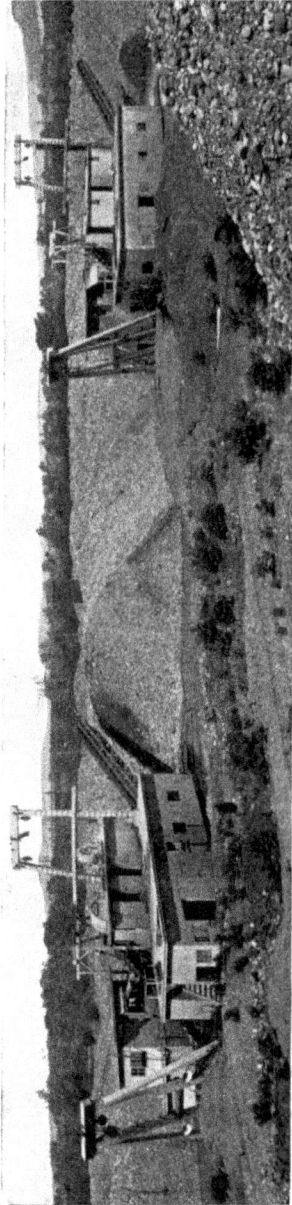

No. 89. Indiana No. 1 and No. 2 dredges, afterwards wrecked by floods.

used at Bannock and Ruby, Montana, as well as in Idaho and California; and the Risdon type of dredge, constructed on the line of those operating in New Zealand, which was the principal type of elevator dredge in use in California at that time.

The operations of the old type Bucyrus and Risdon dredges were considered in a measure satisfactory, and few attempts at improvements

No. 90.   Old sand pump in operation.   Indiana No. 1 Dredge.

had been made, due, probably, to the policy of dredge constructors to discourage the making of any changes, even in minor details, owing to the cost attached thereto, and also because such changes emanated from men considered comparatively inexperienced in this branch of mining. It was, therefore, with difficulty that Messrs. Perry, Griffin, and

No. 91.   Wreck of Indiana No. 2 Dredge in 1907.

Cameron, after failing with other concerns, persuaded the Bucyrus Company, of South Milwaukee, to construct the machinery for a dredge according to their ideas.   The result was the Indiana No. 1 dredge, which as designed by O. B. Perry, F. W. Griffin, D. P. Cameron, and Bucyrus Company, marked a decided change in dredge construction.   It may be said that the large dredges in California, to-day, are improvements along the lines followed in the construction of this dredge.

The Indiana No. 1 dredge was put in commission July 4, 1901, and was in active operation for about six years, during which time it turned over 48.82 acres of ground, and handled 2,645,330 cubic yards of gravel, until destroyed by the floods of March 17, 1907. A general description of this dredge is given in the following: Close-connected-bucket elevator dredge, built to dig 35 feet below the surface of the water, and equipped with plate-girder digging ladder, 78 feet long between centers, carrying 79 close-connected buckets in line, each of 3½-cubic-foot capacity, dumping at a rate of 20 per minute and driven by a 50-horsepower motor. The hull was 86 feet long, 35 feet wide on water-line, and 6 feet 3 inches deep, with a draught of 4 feet 6 inches. The stacker was a Robins belt conveyor, 86 feet long between centers, carrying a 28-inch

No. 92. Indiana No. 3 Dredge, reconstructed from machinery of Indiana No. 2, Oroville District.

belt, 180 feet long, and driven by a 20-horsepower motor. The washing screens were of the flat shaking-screen type, with eccentric drive, and driven by a 20-horsepower motor. An 8-inch centrifugal pump, driven by a 40-horsepower motor, supplied water to the screens and gold-saving tables. The sand pump, which was not in constant use, was driven by a 30-horsepower motor.

The gold-saving tables were constructed of wood and were of the present side-table type, equipped with riffles and quicksilver traps, having a total riffle area of 528 square feet. The collecting and distributing pan used in connection with the washing and gold-saving arrangement, consisted of a steel tray, having the bottom arranged to give a down grade toward the center of the tray, the width and length being equal to the size of the shaking screen. It was placed directly under the screen to collect the screened gravel and to deposit same upon the gold-saving tables placed directly below, through three 3-inch wide, elongated holes, controlled by sliding plate doors.

This dredge was constructed by Griffin and Cameron, who also installed Bucyrus machinery. The satisfactory operation of this dredge brought about the reconstruction of the Continental which was of the Bucyrus double-lift type, equipped with tail sluices and tail scow.

The Indiana No. 2 dredge was put in commission in February, 1903, and constructed slightly larger and heavier than No. 1 dredge, but along the same general lines. It was built to dig 36 feet below the water-line, and was equipped with a plate-girder ladder, 88 feet long between centers, carrying 80 buckets in line, each of 4-cubic-foot capac-

No. 93.   Wreck of Indiana No. 3 Dredge in 1908.   Oroville District.

ity, dumping at a rate of 18 per minute, and driven by a 50-horse-power motor. 133,000 feet of lumber were used in the construction of the hull, which was 80 feet long, 32 feet wide on water-line, and 6 feet 3 inches deep, with a draught of 4 feet. The washing screens were flat shaking with eccentric drive, and the tailing stacker, a Robins belt conveyor. The gold-saving arrangement was on the same order as that installed upon the Continental dredge. The electric motor equipment had a rated capacity of 150-horsepower, distributed as follows: For the supply of water to screens and gold-saving tables, one 30-horsepower, C. S., motor; sand pump, 30-horsepower motor; shaking screen and stacker, one 20-horsepower, C. S., motor; digging or bucket drive motor, 50-horsepower, V. S.; starboard winch motor, 20-horsepower. All the motors were Westinghouse Company, 400 volt, 3 phase, 60 cycles. This dredge was constructed by the Western Engineering and Construction Company, who installed Bucyrus machinery.

Indiana No. 3.—The construction and general design of Indiana No. 3 dredge is practically the same as that of No. 2 dredge.

During the year 1908, the No. 3 dredge handled 250,959 cubic yards of gravel, and up to January 1, 1909, the three dredges handled together a total of about 5,470,370 cubic yards of gravel, and turned over about 100 acres of ground.

The Indiana Company also controls and operates the Indiana Machine Shop Company, in which it employs about twenty-two men a day. In their dredging operations, the company employs about eleven men.

Butte Dredging Company.—Was organized under the laws of the State of Nevada, and the officers are as follows: President and manager, William S. Noyes; secretary and treasorer, B. S. Noyes, Mills Building, San Francisco, California; dredge superintendent, Harrison Appel.

No. 94.   Reconstructing Indiana No. 3, June, 1909.   Oroville District.

Oroville, California. The company owns one dredge which was put in commission November 26, 1902, and dismantled in July, 1910.

The holdings of this company comprise an area of 80 acres located on the west side of the Feather River, in section 7, township 19 north, range 4 east, in Thermalito, and lots 134 to 137 of the townsite of Oroville. Most of the ground was mined by hand or hydraulic processes previous to dredging, and all of it is located on low, bar land, subject to overflow during flood times. The gravel, which is overlain by hydraulic tailing, was prospected by means of drills with test holes at intervals of about one to every four acres. It is a loose river gravel, carrying about 40 per cent of sand, and averaging 34 feet in depth to bedrock.

The Butte dredge is a close-connected-bucket elevator dredge, constructed by the Western Engineering and Construction Company, and equipped with Bucyrus machinery. It was the first dredge to be constructed on the lines laid down in the design of the Indiana No. 1 dredge, but was made larger and heavier than that dredge, the hull being 10 feet longer, and 6 feet wider, and the machinery heavier throughout. It was built to dig 36 feet below water-line. The hull is 90 feet long, 36 feet wide, 7 feet deep, and has a draught of 4½ feet. The digging ladder is box-girder construction, carrying ninety 3½-cubic-foot buckets. The buckets weigh, each, 750 pounds, and the bucket bottoms, 452 pounds. The screens are end-shaking, 7 feet wide, and 25 feet 3 inches long, having an area of 129.6 square feet. The stacker is a Robins belt conveyor, 90 feet long between centers, carrying a 28-inch stacker belt, 188 feet long. The gold-saving tables are arranged on the Holmes system, and have an actual riffle surface of about 750 square feet.

No. 95. Butte No. 1 Dredge, 3½-cubic-foot, California type, Oroville District, July, 1909. Repairing stern. Dismantled in 1910. See pages 205 and 206.

The electric motor equipment installed upon the dredge has a rated capacity of 165-horsepower, distributed as follows:

Pressure pump, 8-inch Worthington centrifugal	40 h.p. motor 440 volts
Primary pump, 2-inch Worthington volute	5 h.p. motor 440 volts
Sand pump, 6-inch Morris sand pump	30 h.p. motor 440 volts
Screen and stacker	20 h.p. motor 440 volts
Bucket drive	50 h.p. motor 440 volts
Winch motor	20 h.p. motor 440 volts
Rated capacity	165 h.p.
Average motor output	86 h.p.

All motors are General Electric Company.

Up to January 1, 1909, or during a little over six years, the dredge turned over 40 acres of ground and handled about 2,665,000 cubic yards of gravel. The company employs an average of eleven men and, taken over a period of five years, has an average monthly pay roll of $1,113.05.

El Oro Dredging Company.—This company began operations in 1904 and up to 1910 has commissioned two dredges, the El Oro No. 1 and No. 2. The company was organized under the laws of Arizona, and the present officers are as follows: President and manager, William S.

Noyes; vice-president, Elwyn W. Stebbins; secretary and treasurer, the company are located in the Mills Building, San Francisco, California.

The company holdings comprise an area of 218.54 acres, located on the east side of the Feather River about one mile inland, in sections 17 and 20, township 19 north, range 4 east. Practically all of this land was covered by placer pits as a result of the early Chinese mining previous to dredging. The gravel deposit is one of the most difficult to dredge in Oroville district; it is very tight and in places partly cemented, and carries throughout considerable clay. The ground was prospected by means of drills, with one test hole to every nine acres.

No. 96.　El Oro No. 1 Dredge.　Constructed by the Link-Belt Company of Chicago, Illinois.

The dredges were designed and constructed by the Link-Belt Company, of Chicago, Illinois, and are of the close-connected-bucket type of placer mining elevator dredge, and were the first Link-Belt Company dredges in the district.

Up to January 1, 1909, while digging to an average depth of 30 feet, the two dredges turned over 67.97 acres of ground and handled 3,310,000 cubic yards of gravel, an average of 84 cubic yards per working hour. During the year 1908, the two dredges handled, together, a total of 1,190,193 cubic yards of gravel.

The El Oro No. 1 dredge was put in commission December 26, 1903, and has been in active operation for over six years. A general description of this dredge is as follows:

Hull.

Length	90　feet
Width	45　feet
Depth	7½　feet
Draught	6　feet

Mechanical Equipment.

Capacity of buckets	5	cubic feet
Average buckets per minute	18	
Number of buckets in chain	90	
Weight of buckets, each	1,533	pounds
Weight of bucket bottoms	928	pounds
Digging ladder, plate-girder type	72 feet 6½	inches long
Screens, end-shaking, area	116.2	square feet
Length of stacker, between centers	90	feet
Stacker belt	30 inches wide, 190	feet long

The gold-saving tables are arranged on the Holmes system, and have a riffle area of 420.7 square feet. The ladder-hoist lines are one inch, the port and starboard bowlines are ¾-inch steel cables. The winch is Link-Belt Company, seven drum. The upper tumbler is pentagon, the lower tumbler is hexagon, and the spuds are made of wood and steel, respectively.

No. 97.   El Oro No. 2 Dredge, Oroville District, California.

The electric motor equipment has a rated capacity of 290-horsepower, distributed as follows:

Pressure pump, 10-inch Worthington centrifugal pump	100 h.p.
Primary pump, Worthington 2½-inch volute pump	5 h.p.
Sand pump, 6-inch Morris sand	40 h.p.
Shaker	20 h.p.
Bucket drive	75 h.p.
Stacker	20 h.p.
Winch	30 h.p.
Rated capacity	290 h.p.

All motors General Electric Company. The average power consumed while in full operation is 137-horsepower or 102 kilowatts.

The El Oro No. 2 was built a little larger and heavier than dredge No. 1, but along the same lines and by the same construction company. This dredge has a total motor capacity of 390-horsepower, an average motor output of 161-horsepower, and an average kilowatt output of 120 kilowatts. It began operations January 22, 1908.

The company employs an average of thirty men, and has an average monthly pay roll of $3,194.93.

Ophir Gold Dredging Company.—This company has an operating plant of two dredges. It was organized under the laws of the State of Nevada. The offices are located in the Clunie Building, San Francisco, California, and the officers are as follows: President, A. F. L. Bell; vice-president, Roger Johnson; secretary and treasurer, F. S. Mayhew; managers, Brayton and Mayhew; superintendent, R. E. Gruber.

The holdings of this company comprise an area of about 211 acres, located in sections 17, 18, and 19, township 19 north, range 4 east, about one mile inland on the east side of the Feather River. Less than one half of this land was planted to orchard, etc., and most of it had been mined previous to dredging. The gravel was well prospected by drills, and averages about 27 feet in depth to bedrock; in character, the deposit is fairly compact, carrying in places considerable clay, but few large boulders.

Part of the holdings of this company belonged formerly to the Nevada Gold Dredging Company, which was originally the Central Gold Dredging Company. The Nevada dredge, now operated by the Ophir Company, was originally commissioned May 15, 1904, by the Central Gold Dredging Company. Later, when this company was taken over by the Nevada Company, the name of the dredge was changed to Nevada and has remained the same after the consolidation of the Nevada Company with the Ophir Company. The Ophir dredge was commissioned August 4, 1906, and is the only dredge put in operation by the Ophir Gold Dredging Company. Both dredges are of the close-connected-bucket elevator type constructed by the Western Engineering and Construction Company, and equipped with Bucyrus machinery.

The total yardage handled by the two dredges during the year ending December 31, 1908, was about 1,064,000 cubic yards of gravel. The total yardage handled by the Ophir dredge during the first three years in operation amounted to about 1,800,000 cubic yards.

The Ophir dredge was built to dig 36 feet below water-line. The hull is 100 feet long, 36 feet wide on water-line, about 40 feet wide on deck line, 7 feet 9 inches deep, and has a draught of 5 feet. The digging ladder is lattice-girder construction 83 feet long between centers, weighs 58,103 pounds, and carries 77 5-cubic-foot buckets, weighing each 1,428 pounds, and dumping at the rate of twenty per minute. The stacker is a Robins belt conveyor, 90 feet long between centers, carrying a 28-inch belt, 186 feet long. The shaking screens are 7 feet and 7 feet 9 inches wide, respectively, and are 14 feet long. The gold-saving tables are of the Holmes system, having an actual riffle surface of 832 square feet.

The electric motor equipment on the dredge has a rated capacity of 262½-horsepower, distributed as follows:

8-inch pressure pump	50	h.p. motor 900 r.p.m.	C.S.
5-inch hopper and primary pump	7½	h.p. motor 1800 r.p.m.	C.S.
Sand pump	50	h.p. motor 900 r.p.m.	C.S.
Shaking screen	20	h.p. motor 1200 r.p.m.	C.S.
Bucket drive	100	h.p. motor 720 r.p.m.	V.S.
Stacker	15	h.p. motor 1200 r.p.m.	C.S.
Winch	20	h.p. motor 1200 r.p.m.	V.S.
Rated capacity	262½	h.p.	

All motors are General Electric Company, 3-phase, 60-cycles, 400-volts.

This dredge was put in operation August 4, 1906, and is in good working condition after a little over three years of active service.

The Nevada dredge was put in commission over two years prior to the Ophir, and is, in every respect, smaller and lighter. This dredge was built to dig 36 feet below the water-line. The hull is 90 feet long, 32 feet wide, on the water-line, 6 feet 3 inches deep and draws 4 feet; the digging ladder is plate-girder construction, and carries 92 4-cubic-foot buckets, weighing, each, 625 pounds, dumping on an average of 19 per minute. The revolving screen is 6 feet in diameter, 24 feet long. The stacker is a Robins belt conveyor, 85 feet long between centers, carrying a 28-inch stacker belt, 176 feet long.

The gold-saving arrangement consists of wood side-tables, equipped with Hungarian riffles and quicksilver traps.

The electric motor equipment installed upon the dredge has a total capacity of 168-horsepower, distributed as follows:

Pressure pump	40 h.p. motor 900 r.p.m.	C.S.
Primary pump	3 h.p. motor 1800 r.p.m.	C.S.
Sand pump	50 h.p. motor 900 r.p.m.	C.S.
Shaking screen	15 h.p. motor 1200 r.p.m.	C.S.
Bucket drive	50 h.p. motor 720 r.p.m.	V.S.
Stacker	15 h.p. motor 1200 r.p.m.	C.S.
Winch	15 h.p. motor 1200 r.p.m.	V.S.
Rated capacity	168 h.p.	

All motors General Electric Company, 440-volt, 3-phase, 60 cycles.

This dredge was put in commission May 15, 1904, and after nearly six years in operation, is in good working condition. It was the first close-connected-bucket elevator dredge to be equipped with revolving screen.

The company employs about twenty men.

Gardella Dredging.—This is a private concern owned by Laurence Gardella, of Oroville, California, and operating one dredge.

The holdings comprise an area of 40 acres, formerly planted to orchard, located in section 18, township 19 north, range 4 east, about one mile inland on the east side of the Feather River.

Gardella dredge, formerly known as California No. 1 dredge, was originally built for the Boston and California Dredging Company in

1903. It was constructed by the Risdon Iron Works and after serving its usefulness with the original company, was sold to Gardella. After working out the Gardella property at Oroville, the dredge was dismantled late in 1909 and moved to Honcut Creek, where it began operations in February, 1910, on the Gardella holdings.

Gold Run Dredging Company began operations in 1906 and has an operating plant of one dredge, known as Baggett No. 1. The officers of this company are as follows: President, N. R. Baggett; first vice-president, R. E. Starr; secretary, D. Jones; treasurer, R. E. Starr; manager,

No. 98. Drilling blast holes in front of dredge for the purpose of loosening gravel. Gold Run property, Oroville District.

L. N. Parks; main office, Mills Building, San Francisco; manager's office, Oroville, California.

The holdings comprise an area of 122 acres, located in sections 20 and 29, township 19 north, range 4 east, on the east side of the Feather River, about one and one fourth miles inland. Practically all of this land was uncultivated and had been mined by hand previous to dredging.

The gravel was prospected by means of shafts, and, like the El Oro deposit, it is very compact and in places partly cemented, carrying considerable clay and some large boulders. The average depth to bedrock is about 35 feet. An electrically driven Keystone drilling machine is operated ahead of the dredge and the ground is loosened by blasting the drill holes.

The Baggett dredge was put in commission May 26, 1906, and since beginning of operations has turned over about 28 acres of ground and handled 1,500,000 cubic yards of gravel. The dredge was designed and constructed by the Risdon Iron Works and is of the new Risdon type. The following is a general description of the dredge:

Hull.

Length	94 feet
Width	34 feet
Depth	7 feet
Draught	4 feet 6 inches

Mechanical Equipment.

Capacity of buckets	7 cubic feet
Buckets per minute	12
Number of buckets in chain	37
Weight of buckets, each	1,500 pounds
Weight of links, each	440 pounds
Digging ladder, solid girder	74 feet long, 45,000 pounds
Revolving screen	26 feet long

Stacker, Risdon bucket type; gold-saving area, 360 square feet.

The electric motor equipment has a total rated capacity of 215-horse-power, distributed as follows:

Motor Equipment.

Pressure pump	50 h.p.
Primary pump	5 h.p.
Revolving screen	10 h.p.
Bucket drive	75 h.p.
Stacker	20 h.p.
Winch	20 h.p.
Ladder hoist	35 h.p.
Rated capacity	215 h.p.

The dredge is roomy, well-kept, and in good working order. Owing to the hard digging it is contemplated changing the bucket-line to close-connected.

The company employs about twelve men.

Viloro Syndicate, Limited.—This company began operations in 1904, and up to 1909, operated one dredge, the Viloro No. 1. The secretary is C. W. Moore, with offices at 5 London Wall Building, London, England. The American agent is H. L. Gunzburger, with offices at 519 California street, San Francisco. The members of the London Committee are Walter McDermott, chairman; Michall Arg David, Thomas H. Leggett. The local manager at Oroville, California, is W. H. James.

The holdings of the company comprise an area of 200 acres, located in sections 5, 19, and 30, township 19 north, range 4 east, on the east side of the Feather River, about one and one half miles inland. The dredgeable area comprises about 170 acres, of which about 150 acres had been previously mined by hand, and none of this ground was suitable for cultivation. The property was prospected by means of drills. The gravel deposit is compact, carrying medium coarse gravel, and in places, considerable clay. The average depth to bedrock is about 30 feet. A little platinum is recovered with the gold.

The Viloro No. 1 dredge was commissioned October 30, 1904, and was destroyed by fire on September 2, 1909, after nearly five years in opera-

No. 99. New type Risdon dredge, showing front gantry with ladder hoist. The Baggett No. 1; 7-cubic-foot buckets. Gold Run Dredging Company, Oroville, Cal.

tion. The disaster occurred during the night, when the hull sprang a leak, and it is thought that when the dredge listed, the electric wires became entangled, thus starting the fire. The dredge had just undergone repairs and had been made ready to resume operations. This company bought the California No. 3 dredge from the Oroville Dredging Limited, to replace the Viloro dredge.

The Viloro No. 1 was a close-connected-bucket elevator dredge, constructed by the Western Engineering and Construction Company, and equipped with Bucyrus machinery. It was built to dig 36 feet below the water-line. The hull was 98 feet long, 36 feet wide on water-line, 7 feet deep, and had a draught of 5 feet. The digging ladder was lattice-girder

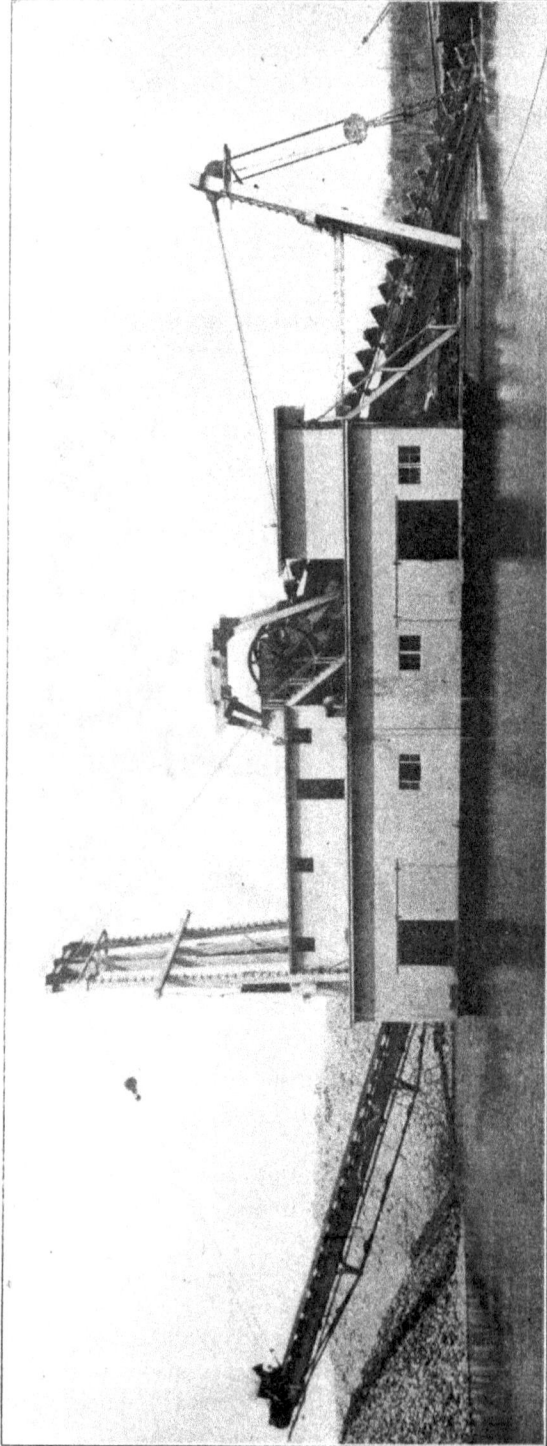

No. 100. Viloro No. 2 Dredge, 7-cubic-foot, formerly California No. 3 Dredge, Oroville District, constructed by the Marion Steam Shovel Company. For description of California No. 3 Dredge, see pages 138 and 139.

construction, 78 feet long between centers, and carried 72 5-cubic-foot buckets, weighing, each, 1,254 pounds, and dumping at the rate of 20 per minute. The shaking screen was 7 feet wide and 28 feet long, with an area of 162 square feet. The stacker was a Robins belt conveyor, 90 feet long between centers, carrying a 30-inch stacker belt, 185 feet long.

The electric motor equipment installed upon the dredge had a rated capacity of 242½-horsepower, distributed as follows: Two 7-inch pressure pumps, 50-horsepower; primary pump, 3-horsepower; sand pump, 30 - horsepower; shaking screen, 20-horsepower; b u c k e t drive, 100-horsepower; stacker, 15-horsepower; winch, 20-horsepower. All motors were General Electric Company, 3-phase, 60-cycles, 440-volt.

The gold - saving arrangement consisted, in a general way, of wood side-tables equipped with Hungarian riffles and quicksilver traps, and had a riffle area of about 1,000 square feet.

Oro Water, Light and Power Company.—This company began operations in 1903 and has an operating plant of five dredges. The officers are, president, J. W. Goodwin; secretary and treasurer, J. K. Moffatt, Chronicle Building, San Francisco, California; manager, Karl Krug, Oroville, California.

The holdings of the company comprise an area of about 1,616 acres, of which about 1,300 are

9—GD

No. 101. Lava Bed No. 1, old type placer-mining dipper dredge. Dismantled.

located on the east side of the Feather River in sections 18, 19, and 30, township 19 north, range 4 east, and in sections 25 and 33, township 19 north, range 3 east, and section 3, township 18 north, range 3 east; and

No. 102. Showing front gantry and bucket-line on an old Risdon dredge. The Marigold No. 2, now out of commission.

the rest on the west side of the river in section 24, township 19 north, range 3 east. The dredgeable area is estimated at about 1,400 acres, all of which is inland, and most was unsuitable for horticultural purposes.

No. 103. Marigold No. 2 Dredge in 1909. Oroville District.

The company took over the holdings of the Lava Bed Dredging Company with an operating plant of two new elevator dredges, and one dredge out of commission; and the Marigold Gold Dredging Company with two Risdon elevator dredges, now out of commission. Three of the dredges now in operation were installed by the present company and are all of the close-connected-bucket elevator type.

The names of the dredges at present operating are as follows: Lava Bed No. 2, and No. 3, Empire, Hunter, and Victor. All of these dredges are 5-cubic-foot close-connected-bucket elevator dredges, constructed by the Western Engineering and Construction Company and equipped with Bucyrus machinery. Mechanically they are practically the same. A general description of the dredges is as follows:

Lava Bed No. 2.—This dredge, which was originally constructed for the Lava Bed Dredging Company, was put in commission in July, 1903. It was built to dig 36 feet below the water-line, and is equipped with 5-cubic-foot buckets. 190,000 feet of lumber were used in the construction of the hull, which is 96 feet long, 36 feet wide on water-line, and 7 feet deep, with a draught of 4 feet 6 inches. The stacker is a Robins belt conveyor, 90 feet long between centers, carrying a 30-inch belt, 187 feet long. The washing screens are flat, shaking with eccentric drive, and the gold-saving tables are the same as those ordinarily installed upon standard California elevator dredges.

The electric motor equipment installed upon the dredge has a rated capacity of 208-horsepower, distributed as follows:

Pressure pump, for supply of water to screens and gold tables	50 h.p. motor 850 r.p.m.	C.S.
Primary pump, for supply of water to hopper	3 h.p. motor 1700 r.p.m.	C.S.
Sand pump, not often used	30 h.p. motor 850 r.p.m.	C.S.
Shaking screen and tailing stacker motor, together	30 h.p. motor 850 r.p.m.	C.S.
Main digging or bucket drive motor	75 h.p. motor 600 r.p.m.	V.S.
Starboard winch motor	20 h.p. motor 1200 r.p.m.	V.S.
Rated capacity	208 h.p.	

All motors are Westinghouse Company, 3-phase, 60-cycles, 400-volt.

Lava Bed No. 3.—This dredge, which was constructed for the Lava Bed Dredging Company, was put in commission December 4, 1904. It was constructed slightly larger and heavier than No. 2 dredge, but along the same general lines. It was built to dig 40 feet below the water-line, and is equipped with 5-cubic-foot buckets, which dump at the rate of 20 per minute. Each of the buckets weighs 1,125 pounds, and the plate-girder digging ladder has a total weight of 54,524 pounds. The washing screens are flat, shaking, and the gold-saving tables are arranged much on the same order as those used on No. 2 dredge. The tailing stacker is a Robins belt conveyor, 90 feet long, carrying a 30-inch belt, 187 feet long. It required 195,000 feet of lumber in the construction of the hull, which is 104 feet long, 36 feet wide on water-line, and 7 feet deep, with a draught of 4 feet 6 inches.

The electric motor equipment as installed upon the dredge has a rated capacity of 203-horsepower, distributed as follows:

Pressure pump, two 7-inch direct-connected to
  one _____ 50 h.p. motor 685 r.p.m. C.S.
Primary pump _____ 3 h.p. motor 1800 r.p.m. C.S.
Shaking screen _____ 15 h.p. motor 1200 r.p.m. C.S.
Main digging or bucket drive_____ 100 h.p. motor 600 r.p.m. V.S.
Tailing stacker _____ 15 h.p. motor 1200 r.p.m. C.S.
Starboard winch _____ 20 h.p. motor 900 r.p.m. V.S.

    Rated capacity _____ 203 h.p.

All motors are Westinghouse Company, 3-phase, 60-cycles, 400-volt.

Empire Dredge.—This dredge was put in commission April 22, 1906. It was built to dig 38 feet below the water-line, and is equipped with 82 5-cubic-foot buckets weighing, each, 1,330 pounds, and dumping at the rate of 19 per minute. The digging ladder weighs complete

No. 104.   Empire Dredge, 5-cubic-foot, California type.   Oroville District.

63,925 pounds. The revolving screen is 24 feet long and 6 feet in diameter; and the tailing stacker is a Robins belt conveyor, of lattice-girder construction, 90 feet long between centers, and carries a belt 30 inches wide and 187 feet long. It required 210,000 feet of lumber in the construction of the hull, which is 102 feet long, 36 feet wide on water-line, and 7 feet 9 inches deep, with a draught of 5 feet. For supplying water to screen and gold tables, which are of the Holmes type, the dredge is equipped with one 8-inch centrifugal pump, delivering 1,800 gallons per minute against a 60-foot head. Water for the hopper and spraying pumps is being supplied by a 4-inch centrifugal pump, operating against a 65-foot head.

The electric motor equipment as installed upon the dredge has a rated capacity of 225-horsepower, distributed as follows:

Pressure pump, one 8-inch centrifugal_____ 50 h.p. motor 850 r.p.m. C.S.
Primary pump, one 4-inch_____ 15 h.p. motor 1120 r.p.m. C.S.
Revolving screen and stacker, together_____ 20 h.p. motor 1120 r.p.m. C.S.
Main digging or bucket drive_____ 100 h.p. motor 600 r.p.m. V.S.
Tailing stacker _____ 20 h.p. motor 1120 r.p.m. C.S.
Starboard winch _____ 20 h.p. motor 900 r.p.m. V.S.

    Rated capacity _____ 225 h.p.

Hunter Dredge.—This dredge was put in commission August 13, 1907. It is built to dig 38 feet below the water-line, and is equipped with 82 5-cubic-foot buckets, dumping at the rate of 20 per minute, a n d weighing, each, 1,294 pounds. It required 215,000 feet of lumber in the construction of the hull, which is 102 feet long, 36 feet wide, on water-line, and 7 feet 9 inches deep, with a draught of 5 feet. The revolving screen is 25 feet 6 inches long, with a diameter of 6 feet, and operates at a speed of nine revolutions per minute. The stacker is a Robins belt conveyor, of lattice - girder construction, 102 feet long between centers, carrying a belt 30 inches wide, and about 212 feet long. The gold-saving tables are the same as those installed upon the standard elevator dredges in California. The water supply for washing a n d sluicing purposes is furnished by o n e 8-inch centrifugal pump, delivering 1,800 g a l l o n s p e r minute against a 50-foot head, and the water supply for h o p p e r a n d priming pumps, etc., is furnished by one 4-inch centrifugal pump, operating against

No. 105. Victor Dredge, 5-cubic-foot. California type. Oroville District. For illustration of Hunter dredge, see page 114.

a 65-foot head.   The electric motor equipment as installed upon the dredge has a rated capacity of 225-horsepower, distributed as follows:

Pressure pump, one 8-inch centrifugal_____	50 h.p. motor 850 r.p.m.	C.S.
Primary pump, one 4-inch centrifugal_____	15 h.p. motor 1120 r.p.m.	C.S.
Revolving screen _____	20 h.p. motor 1120 r.p.m.	C.S.
Main digging or bucket drive, one_____	100 h.p. motor 600 r.p.m.	V.S.
Tailing stacker, one _____	20 h.p. motor 1120 r.p.m.	C.S.
Starboard winch motor, one_____	20 h.p. motor 900 r.p.m.	V.S.
Rated capacity _____	225 h.p.	

All the motors are Westinghouse Company 3-phase, 60-cycle, 400-volt.

The Victor dredge was put in commission September 4, 1907, and in construction is practically a duplicate of the Hunter dredge.   The company employs an average of 60 men.

New York Machine Shop.—Aside from dredging operations, the Oro Water, Light, and Power Company controls and operates the New York Machine Shop, which is well equipped and large enough to handle all the repairs necessary to the dredges.

Pennsylvania Gold Dredging Company.—This company began operations in 1902, and has an operating plant of one dredge.   The company is controlled by S. W. Cheney, with offices at 327 First street, San Francisco.   James Nesbit is superintendent at Oroville.

The company holdings comprise an area of 155.5 acres, located on the east side of the Feather River in section 25, township 19 north, range 3 east, and formerly the property of the California Dredging Company, which had an operating plant consisting of one dipper dredge.   Most of the ground had been mined by hand previous to dredging and none of it was cultivated with success.   The property was prospected by means of drills.   The gravel is medium coarse and free from clay, averaging about 28 feet in depth to bedrock.

The Pennsylvania dredge was put in operation in November, 1902. It was designed and constructed by the Golden State Miners' Iron Works of San Francisco, and differs in many points of construction from those operating in the field.   It is of the close-connected-bucket elevator type, having 54 buckets in line, each of 6-cubic-foot capacity, and made of solid cast steel, each weighing 1,450 pounds.   The hull is 80 feet long, and 36 feet wide on water-line, with a draught of $5\frac{1}{2}$ feet.   The gold-saving arrangement is of special design, having a riffle surface of about 800 square feet.   There are two shaking-screens, one above the other; the upper having $1\frac{1}{2}$-inch holes, and the lower, $\frac{1}{4}$-inch holes.   After the upper screen wore out it was not replaced, so that the dredge has, now, but one set of shaking-screens, similar to those in use on the other dredges.   The stacker is a belt conveyor, driven from the outer end.

The electric motor equipment installed upon the dredge has a total capacity of 220-horsepower, distributed as follows: Pressure pump and shaking screen motor 75-horsepower, rated capacity, consuming 40-horsepower while in full operation; bucket drive motor, 75-horsepower; stacker drive motor, 10-horsepower; winch motor, 20-horsepower; river pump motor, 30-horsepower; and one extra pump motor, 10-horsepower.

Up to January 1, 1909, the dredge has handled about 3,500,000 cubic yards of gravel. See illustration No. 107 on page 136.

Oroville Dredging, Limited.—This company has an operating plant of six dredges and one machine shop. The company is capitalized for $3,500,000, divided into 700,000 shares of $5 each. The officers are as follows: Directors, chairman, Frederick William Baker, 3 and 4 Lothbury, London, E. C.; Francis David Behrend, 3 Little Stanhope street, London W.; Sidney Arthur Bird, 3 and 4 Lothbury, London, E. C.; Henry David Boyle, 8 Old Jewry, London, E. C.; secretary, Henry Richards, 5 Moorgate street, London; general manager,

No. 106. Old type dipper dredge, the California No. 1, now abandoned. Pennsylvania Dredging Company.

No. 107. The Pennsylvania Dredge. Golden State Miners' Iron Works construction.

W. P. Hammon, Alaska Commercial building, San Francisco, California; local manager at Oroville, John J. Hamlyn.

In the Oroville district the holdings of the company comprise an area of about 1,346 acres, located in sections 18 and 19, township 19 north, range 4 east, on the east side of the Feather River, extending as far as one mile inland, and in sections 23, 24, 25, and 26, township 19 north, range 3 east, on the west side of the Feather River, extending as far as two miles inland. With the exception of about 200 acres, none of this land has been under cultivation, and practically all of it was mined previous to dredging. The properties were carefully prospected by means of drills. The gravel averages about 12½ cents per cubic yard, and in character ranges from a clean, loose, river wash, to a heavy bench gravel, in places partly cemented, and at times carrying considerable clay. The average depth to bedrock is about 30 feet.

In the Bear River district, the company owns 993 acres, located in the Bear River Basin in Placer and Yuba counties.

It being the policy of the management of the Oroville Dredging, Limited, to preserve the identity of each of the companies operating under the consolidation, a general description of the operations of these subsidiary companies is given. The following is a general summary of the operations for the twelve months ending July 31, 1907, during which time there were twelve dredges in operation, a greater number than at any time since the consolidation.

	Total Cost.	Cost per Cubic Yard in Cents.	Per Cent of Total Cost.
Operating	$116,402.56	1.46	27.3
Electricity	57,013.33	.71	13.4
Water	6,995.00	.09	1.7
Repairs	184,509.61	2.44	45.7
General expenses	34,997.06	.44	8.2
Taxes and insurance	15,904.14	.20	3.7
Total expense	$415,821.70	5.34	100.

Gross output, $895,024.92; gross output per cubic yard, 11.23 cents. Net output, $479,203.22; net output per cubic yard, 5.89 cents.

Companies comprising the Oroville Dredging, Limited:

Property.	Dredges put in Commission	Dredges Operating in 1910.	Acres Owned.
Boston and California Dredging Co.	3	1	271
Boston and Oroville Mining Co.	4	1	469
Oroville Gold Dredging and Exploration Co.	3	3	606
Oroville District	10	5	1,346
Bear River Mining Co.	5	1	993.10
	15	6	2,339.10

Actual dredging time, 73,856 hours 30 minutes; average dredging time, daily, 17 hours 54 minutes; cubic yards excavated, 7,793,678; average yardage daily, 1,923 cubic yards; acres dredged, 163.28; average depth of ground, 30.3 feet, year ending July 31, 1907. See pages 85 to 102 for other working costs of this company.

The Boston and California Dredging Company began operations in 1902. It took over the holdings of the Leggett and Wilcox Company, which had an operating plant consisting of two dredges, known as Leggett No. 1 and No. 2. Part of the holdings, the Wilcox tract, was land owned by the company, the Leggett tract being land leased by the company on a royalty basis. Both of these companies were promoted by W. P. Hammon, and operated under his direction, the personnel consisting, mainly, of Boston men. After the consolidation, the dredges were named California No. 1 and No. 2.

California No 1 dredge was a 5-cubic-foot open-link-bucket elevator dredge constructed by the Risdon Iron Works, and was one of the largest dredges of its kind at the time. After some time in operation it was remodeled by the operating company and equipped with a close-connected Bucyrus bucket-line and 3½-cubic-foot buckets. While equipped with this bucket-line, during ten months in 1906 and 1907, this dredge handled 394,156 cubic yards of gravel, at a cost of 6.23 cents per cubic yard. After serving its usefulness with the company, the dredge was sold to L. Gardella.

California No. 2 dredge was put in commission in December, 1902. It is a 5-cubic-foot close-connected-bucket elevator dredge, equipped with Bucyrus machinery. This was the fourth dredge to be constructed by the Western Engineering and Construction Company on the lines of Indiana No. 1 dredge. It was built to dig 36 feet below water-level. The size of the hull is 96 feet long, 36 feet wide, and 7 feet deep, with a draught of 4 feet 6 inches. The bucket ladder is plate-girder construction; the screens are end-shaking; the stacker, Robins belt conveyor, 90 feet long between centers, carrying a 30-inch belt. The gold-saving arrangement consists of wood side-tables, equipped with Hungarian riffles and quicksilver traps. The electric motor equipment installed upon the dredge has a rated capacity of 195-horsepower, distributed as follows: Two 6-inch pumps to supply water to the screen, gold tables, and spray to hopper, direct-connected to a 40-horsepower motor; one 6-inch sand pump, 30-horsepower; shaking screen and stacker motor, 30-horsepower; bucket-drive motor, 75-horsepower; starboard winch, 20-horsepower. All motors are Westinghouse Company. 60-cycles, 3-phase, 400-volt.

California No. 3 dredge was put in commission in October, 1904. It is a 7-cubic-foot close-connected-bucket elevator dredge, constructed by the Marion Steam Shovel Company, and designed by the Boston Machine Shop Company and the Marion Company.

This dredge has handled over 1,100,000 cubic yards of gravel in twelve months' time, in fairly difficult ground, and has a capacity of 120,000 cubic yards per month, while operating in free, loose, gravel. It was in constant operation up to the end of 1909, when it was sold to the Viloro Syndicate, after turning over the property it was working on. It is intended to have the dredge dig its way to the Viloro property, where it will replace the Viloro dredge, which was destroyed by fire.

The California No. 3 dredge was one of the largest and most efficient dredges in the field at the time of construction, many improvements and new features being brought out, several of which are worthy of mention. The ladder-hoist is provided with an automatic, friction, brake-wheel, which gives easy and uniform control in raising and lowering the ladder. A slipping friction is placed inside the pulley of the motor shaft to compensate sudden stalling of power as it is a protection for the motor. The stacker hoist is so arranged to enable the stacker to be raised and lowered by means of a reversible worm-screw, meshing into a gear, and counter-shafted to a cable drum, thus locking it at any given point. Much attention was given to the design and shape of the buckets. They were made of 7-cubic-foot capacity, with a 34-inch pitch, the hood being made of one piece of steel plate. The bucket-ladder is plate-girder construction, 80 feet long, weighing 60,000 pounds, and carries 68 buckets in line, each weighing 1,800 pounds. The tailing conveyor is 92 feet 6 inches long, weighs 34,000 pounds, and is equipped with flat rollers and flanged idlers at each end. The conveyor belt is 32 inches wide, and in practice has shown great endurance by reason of being carried by flat rollers. The cylinder drums at the lower and upper ends of stacker are 36 inches in diameter. The belt is driven by a motor placed at the outer end of the stacker. The revolving screen is of the C. W. Gardner stepped type, the shoulders of which hold the material sufficiently in suspension to thoroughly wash it. A distributing V-shaped casing is placed underneath the screen with two gates to each section of gold-saving tables and deflecting plates are used at the opening of these gates to evenly distribute the material over the tables. The gold-saving tables have an area of 648 square feet, and a total riffle surface of 894 square feet, divided into nine sections, each 2 feet 8 inches wide, placed on each side of the screen. See California No. 3 dredge on page 128.

The Boston and Oroville Mining Company began operations in 1901. This company was among the first started by W. P. Hammon. It had at one time, an operating plant of four dredges, but at the present has only one dredge in active operation.

No. 108. The Boston No. 1 dredge, Risdon type, now dismantled.

The Boston No. 1 dredge began operations in July, 1901. It was a 5-cubic-foot open-link-bucket elevator dredge constructed by the Risdon Iron Works. After some years in operation it was remodeled by the Boston Machine Shop Company and equipped with a close-connected-bucket-line and 3-cubic-foot buckets.

During the twelve months ending July 31, 1907, this dredge handled, with the new bucket-line, 479,473 cubic yards of gravel, at a cost of 6.58 cents per cubic yard, while digging to an average depth of 32.3 feet. In 1908 the dredge was put out of commission, owing to the gravel becoming too tight for this dredge to economically handle.

The Boston No. 2 dredge was put in commission in July, 1902. It was of the same make and design as No. 1 dredge, and was also remodeled by the operating company, and equipped with a close-connected-bucket-line, and 3-cubic-foot buckets. In 1908 this dredge was permanently put out of commission.

The Continental dredge began operations in 1899. This was the first electrically operated dredge in California, the first on which variable speed motors were used, as well as the first gold dredge to be equipped with close-connected buckets. It was originally built as a pump suction double-lift elevator dredge equipped with tail scow, tail sluices and open-link buckets. F. W. Griffin and Ben Stanly Revett, who owned the Continental property, interested Boston people, who have since been prominent in dredging on the Continental ground and elsewhere in the Oroville District. In 1901 the dredge was reconstructed by the original constructors, Griffin and Cameron, and converted into a 4-cubic-foot close-connected-bucket elevator dredge of the present standard design. Among the principal changes made was the rearranging of the gold-saving device from straight sluices to riffle tables, and to enable clean washing and even distribution of the material, a set of shaking screens was installed, together with a distributing pan, placed under the screens. The distributing pan used upon this dredge was changed and im-

No. 109. The Continental dredge after reconstruction. Rock chute on side not used on up-to-date boats.

proved over the one installed on Indiana No. 1. This improved distributor was designed by D. P. Cameron, and was provided with a deep collecting trough in the center of the pan, having a series of openings on each side to feed the gravel directly into each sluice of the gold-saving tables. An arrangement of four distributing troughs was installed on the bottom of the pan on each side of the center line to more evenly distribute the gravel throughout the length of the distributing pan, which deposits the gravel upon the riffle sluices of the gold-saving tables. This improved distributor, with its center gathering trough and separate side openings to each center gathering trough and separate side openings to each sluice is much the same as those being used to-day on the latest dredges equipped with sluicing screens. This dredge was in almost constant operation for ten years and four months, until permanently shut down in November, 1909, after having worked out the property.

Boston No. 4 dredge was put in commission in May, 1906. It is a 7-cubic-foot close-connected-bucket elevator dredge, constructed by the Boston Machine Shop Company, and equipped with Marion machinery. During the first year in operation it handled an average of 93,000 cubic yards of gravel per month, while digging to an average depth of 28 feet. It is the only dredge being operated by the company at the present time.

In the construction of the Boston No. 4 dredge, an improvement was made in the construction of the forward gantry frame. In place of the old "A" frame design, four parallel upright posts, strengthened with braces are securely tied to the main framing of the boat, and slope forward at an angle of 5 1-16 inches horizontal to 1 foot vertical. There are 68 close-connected buckets in line, dumping at a rate of 20 per minute and driven by a 150-horsepower motor. The upper tumbler is pentagonal, and the lower hexagonal. The tailing stacker is a belt conveyor; the revolving screen is 30 feet long and 6 feet in diameter. The spuds are, as usual, one wood and one steel. The motor equipment installed upon the dredge has a rated capacity of 360-horsepower, divided as follows: High pressure pump, 50-horsepower; low pressure pump, 25-horsepower; main digging motor and ladder hoist, 150-horsepower; priming pump, 10-horsepower; sand pump, 75-horsepower; winches, 25-horsepower; screen and stacker drive, 25-horsepower.

Oroville Dredging and Exploration Company began operations in 1902. The personnel of the company consisted of John Hays Hammond, F. W. Bradley, J. E. Doolittle, and others. The holdings are located on the west side of the Feather River.

Exploration No. 1 dredge, formerly known as Biggs No. 1, was put in commission in April, 1902. It was a 4-cubic-foot open-link-bucket elevator dredge, constructed by the Risdon Iron Works, and later reconstructed by the present operating company, and equipped with a close-connected bucket line, and buckets of 3-cubic-foot capacity.

This dredge is an example of the earlier Risdon boats, and was constructed with the pitched roof of the housing and the gold-saving tables and trommel located outside the housing. The hull of this dredge is 86 feet long, 30 feet wide and 7 feet deep. The digging ladder carries 78 close-connected buckets, each of 3-cubic-foot capacity. The washing screen is 4 feet 6 inches in diameter and 25 feet long. The 12 sections of the gold-saving tables slope from the screen towards each side, and empty into side sluices. The tailing stacker is of the usual Risdon bucket type, and headlines are used in place of spuds.

Exploration No. 2 dredge, formerly known as Biggs No. 2, began operations in 1905. It is 5-cubic-foot close-connected-bucket elevator dredge, constructed by the Western Engineering Construction Company, and equipped with Bucyrus machinery, and was built to dig 30 feet below water-level. The hull is 94 feet long, 36 feet wide, 7 feet deep, and draws 4 feet 6 inches. The digging ladder is lattice-girder construction and weighs 45,775 pounds, including rollers and bearings, etc. The screens are end-shaking, 14 feet wide and 26 feet long. The stacker is lattice-girder construction, 90 feet between centers, carrying a belt 30 inches wide. The gold-saving

No. 110. Exploration No. 2, formerly known as Biggs No. 2 Dredge. 5-cubic-foot buckets.

area consists of wood side-tables, equipped with Hungarian riffles and quicksilver traps.

The electric motor equipment installed upon the dredge has a rated capacity of 238-horsepower, divided as follows: One 50-horsepower motor, direct-connected to two 7-inch pumps for the supply of water to screen; one 30-horsepower motor, connected to a 6-inch sand pump; shaking screen motor, 20-horsepower; bucket-drive motor, 100-horsepower; starboard winch motor, 20-horsepower; tailing stacker, 15-horsepower; priming pump, 3-horsepower. This dredge is one of the three now being operated by this company.

No. 111.   Showing top soil and part of dredge pond, Oroville, Cal.

Exploration No. 3 dredge was put in commission October 20, 1906. It is a close-connected-bucket elevator dredge, constructed by the Boston Machine Shop Company, and equipped with Marion machinery and 7-cubic-foot buckets. During the first ten months in operation, this dredge handled an average of 90,000 cubic yards of gravel per month, while digging to an average depth of 35½ feet in difficult ground.

The Boston Machine Shop Company was organized for the purpose of centralizing the purchase of supplies for the different dredges operated under Oroville Dredging, Limited, thereby securing the rebates recruing from the purchase of large lists of dredge parts, and also for the purpose of constructing and repairing the machines without the loss of time incident to the placing of these orders at distant points.

The Feather River division of the Natomas Consolidated of California has an operating plant consisting of two dredges. On January 1,

1909, the Natomas Consolidated of California took over the holdings of the Cherokee Gold Dredging Company, having an area of 200 acres, and an operating plant of one dredge; and the Feather River Exploration Company, having an area of about 900 acres and five dredges, some of which were abandoned.

The total holdings of the company comprise an area of about 1,364 acres, located in sections 24, 25, and 26, township 19 north, range 3 east, on the west side of the Feather River, extending as far as one mile inland; and in section 3, township 18 north, range 3 east, and section 24, township 19 north, range 3 east, on the east side of the river, extending

No. 112. The Cherokee Dredge, July, 1909, California type working on headline. Oroville District. Closed down December 31, 1909.

as far as one and a half miles inland. With the exception of about 80 acres, none of this land has been under cultivation, and most of it was mined by hand previous to dredging. On some of this land the first dredging operations in the Oroville district were carried on. The gravel averages from 9 cents to 14 cents per cubic yard, the upper part of the property being the richer, and in character the gravel is a clean river wash, in places overlain by several feet of heavy loam. The average depth to bedrock ranges from 25 to 40 feet.

Feather No. 1 dredge, formerly known as the Cherokee, was originally built for the Cherokee Gold Dredging Company and put in commission in July of 1902. It operated for over seven years and was in good working condition when shut down December 31, 1909.

The Cherokee is a close-connected-bucket elevator dredge, built to dig 45 feet below the water-line, and equipped with 5-cubic-foot

10—GD

buckets. The hull is 93 feet long, 36 feet beam, 7 feet deep, with a draught of 4 feet 6 inches. About 200,000 feet of lumber were used in the construction of the hull. The stacker is a Robins belt conveyor, 90 feet long between centers, carrying a 30-inch belt. The ladder is plate-girder construction, and the gold-saving tables are of the wood side-table type, equipped with Hungarian riffles and quicksilver traps. The washing screens are end-shaking.

This dredge differed from the usual type of Bucyrus dredges in that it worked on a headline, in place of spuds. It was constructed by Griffin and Cameron, and equipped with Bucyrus machinery.

The electric motor equipment installed upon the dredge has a rated capacity of 205-horsepower, and the average power consumed while in full operation is about 203-horsepower.

No. 113. Feather River No. 2 Dredge. Risdon type. Not in operation.

The Feather River Exploration Company was the first company to begin dredging operations in the Oroville district. It was originally started by W. P. Hammon and Captain Thomas Couch, and to these enterprising men much credit is due for the rapid development of gold dredging in California since 1897. The holdings of this company were taken over by the Feather River Development Company in January, 1906.

The first dredge to be built by the Feather River Exploration Company was known as Couch No. 1, and later operated as Feather River No. 1 dredge. It was put in commission March 1, 1898, and was an early Risdon machine, constructed on the lines of those used in New Zealand. It was the first successful bucket elevator dredge in California.

The next two dredges to be put in operation by the Feather River Development Company were known as Couch No. 2 and No. 3, and later operated as Feather River No. 2 and No. 3 dredges. They were put in commission on June 8, 1900, and operated successfully until shut down. Like No. 1 dredge they were both designed and constructed by the Risdon Iron Works.

The fourth dredge to be built by the Feather River Development Company was known as Couch No. 4, and later operated as Feather River No. 4 dredge, and when the Feather River Exploration Company

No. 114.   Feather No. 4 Dredge, July, 1909, formerly known as the Feather River No. 5.
Risdon type.   Oroville District.   Closed down December 31, 1909.

was taken over by the Natomas Consolidated of California, this dredge was known as Feather No. 4.   It was constructed by the Risdon Iron Works on the same lines as the other Couch dredges, and was put in

No. 115.   Feather No. 2 Dredge, showing extra gold-saving tables on outside of housing.

commission December 10, 1902, and did good work until closed down December 31, 1909.

The fifth of the Feather River dredges to be put in operation was built for the Feather River Exploration Company by the Risdon Iron Works, and put in commission in January, 1903.   It was wrecked by the floods in 1907, and subsequently dismantled. It operated under the name of Feather River No. 5 dredge.

Feather No. 2 and No. 3.—The only dredges now being operated in the Oroville district by the Natomas Consolidated of California are the Feather No. 2 and No. 3 dredges, which were put in commission on December 22, 1906, and March 26, 1908, respectively. They are both close-connected-bucket elevator dredges, constructed by the Yuba Construction Company, and equipped with 7½-cubic-foot buckets and Marion machinery.

The Natomas Consolidated of California contemplates the building of a 13½-cubic-foot dredge during the year 1910, in the Oroville district. This company employs about forty-five men.

No. 116.   Gravel and sand bank in front of dredge, Oroville.

Pacific Gold Dredging Company.—This company began operations May 1, 1906, and has an operating plant of four dredges. It is a subsidiary company to the Yukon Gold Company, the officers being as follows: President, S. R. Guggenheim; general manager, O. B. Perry; secretary, Chas. K. Lipman, 165 Broadway, New York; manager, O. C. Perry, Oroville, California.

The holdings comprise an area of 750 acres, located in sections 2, 3, and 4, township 18 north, range 3 east, and sections 33 and 34, township 19 north, range 3 east, on the east side of the Feather River, extending as far as 1½ miles inland. With the exception of about 100 acres, none of this land has been under cultivation, and a great deal of it was mined by hand previous to dredging. The gravel was well prospected by means of drills, and averages in depth to bedrock about 28 feet, and yields an average of about 9 cents per cubic yard; in character it is mostly a clean, loose gravel, in places overlain by fine loam and sand to a depth of several feet.

This company took over the holdings of the American Gold Dredging Company, having an operating plant of two dredges, and an acreage

No. 117.   The Kia Ora Dredge in operation, old Risdon type.   Dismantled.

of 275 acres; and the Kia Ora Gold Dredging Company with an acreage of 153 acres and one abandoned dredge.   In the following a general description is given of the dredges being operated by the company:

Pacific No. 1 was put in operation May 1, 1906, and when originally built was known as Perry No. 1 dredge.   During the first two years and

No. 118.   Kia Ora Dredge in 1909.   Oroville District.   Old Risdon type.

No. 119. Pacific No. 1 Dredge, 7-cubic-foot. California type. Oroville District. Western Engineering and Construction Company design.

eight months in operation it turned over 64.75 acres of gravel, averaging 34½ feet in depth.

It is a close-connected-bucket elevator dredge, constructed by the Western Engineering and Construction Company, and equipped with Bucyrus machinery. The hull is 95 feet long, 38 feet 6 inches wide, 8 feet 3 inches deep, and draws 5 feet 9 inches. The digging ladder is lattice-girder construction, and carries 84 buckets of 7½-cubic-foot capacity each, dumping 20 per minute. The screens are end-shaking; the stacker is lattice-girder construction, 90 feet long between centers. The gold-saving tables are modified Holmes type. The electric motor equipment installed upon the dredge has a rated capacity of 300-horsepower, distributed as follows: For the supply of water to the screen and tables, one 50-horsepower motor connected to one 10-inch pump, and one 25-horsepower motor connected to one 8-inch pump; for the supply of water to spraying hopper, one 15-horsepower motor connected to a 4-inch pump; shaking screen motor, 20-horsepower; sand pump motor, 50-horsepower; tailing stacker, 20-horsepower motor; winch motor, 20-horsepower; and digging or bucket drive motor, 100-horsepower.

During the twelve months ending January 1, 1909, the dredge handled an average of 117,205 cubic yards per month, at a cost of 4.44 cents per cubic yard, including dredge depreciation charges, and 3.76 cents per cubic yard, exclusive of dredge depreciation charges. The total number of working hours during the year were 6,817.

Pacific No. 2 dredge began operations in 1902. It is a 4-cubic-foot close-connected-bucket elevator dredge, originally constructed by Griffin & Cameron for the American Gold Dredging Company, and then known as the American No. 1 dredge.

Pacific No. 3 dredge began operations in April, 1904. It was originally constructed for the American Gold Dredging Company, and was known as the American No. 2. It is also a close-connected-bucket elevator dredge, equipped with 5-cubic-foot buckets, and constructed by the Western Engineering and Construction Company. It is equipped with Bucyrus machinery, and was built to dig 36 feet below water-level. The hull is 96 feet long, 36 feet wide, 7 feet deep, with a draught of 4½ feet. The ladder is plate-girder construction and the stacker a Robins' belt conveyor, 90 feet long between centers, carrying a 30-inch belt. The tables are modified Holmes type. The electric motor equipment installed upon the dredge has a rated capacity of 208-horsepower, distributed as follows: Main pressure pump, 50-horsepower; primary or hopper pump, 3-horsepower; sand pump, 30-horsepower; shaking screen motor, 15-horsepower; bucket drive motor, 75-horsepower; stacker motor, 15-horsepower; winch motor, 20-horsepower. This dredge is conveniently arranged and well kept.

Pacific No. 4 began operations in January, 1908, and during the first twelve months in operation this dredge handled 132,452 cubic yards of gravel, and turned over 42.04 acres, while digging to an average depth of 28 feet 10 inches. The average cost per cubic yard was 5 cents, including dredge depreciation charges, and 3.76 cents exclusive of dredge depreciation charges. The total number of working hours during the year were 7,161. The gravel was fairly fine, carrying a large proportion of sand.

The Pacific No. 4 dredge is a 7-cubic-foot close-connected-bucket elevator dredge, constructed by the Western Engineering and Con-

No. 120. Gold-saving tables on Pacific No. 4 Dredge, Oroville, Cal.

struction Company, and equipped with Bucyrus machinery. It was built to dig 36 feet below water-line. The hull is 95 feet long, 38 feet 6 inches wide, 8 feet 4 inches deep, with a draught of 6 feet. The total amount of lumber in hull is 230,000 feet. The digging ladder is plate-girder, weighs 78,685 pounds complete, and carries 80 buckets of 7-cubic-foot capacity, weighing, each, 1,777 pounds, dumping 19 per minute. The screens are end-shaking, with eccentric drive. The upper screen is 7 feet 11 inches wide, 14 feet long, and the lower 8 feet 8 inches wide and 14 feet long. The tailing conveyor is lattice-girder, 90 feet long between centers, carrying a 32-inch belt. The electric motor equipment installed upon the dredge has a rated capacity of 300-horsepower, distributed as follows: For the supply of water to the screens and gold-saving tables, one 75-horsepower motor connected to one 8-inch and one 10-inch pump. For the supply of water to hopper, one 15-horsepower motor direct-connected to a 4-inch pump; and one 8-inch sand pump motor, 50-horsepower; shaking screen motor, 20-horsepower; bucket drive motor, 100-horsepower; stacker motor, 20-horsepower; winch motor, 20-horsepower.

The company employs forty-two men, at an average wage of about $3 per day, including dredgemaster.

Leggett Gold Dredging Company.—This company began operations

No. 121.   Leggett No. 3 Dredge, 5-cubic-foot, Risdon type.
Now dismantled.   Oroville District.

in March, 1904, and had an operating plant of one dredge; it was a close corporation, with James H. Leggett president.

The holdings comprised an area of about 75 acres located in section 18, township 19 north, range 4 east, on the east side of the Feather River, about one half mile inland. The land was part of the Leggett ranch, all planted to orchard previous to dredging.

The Leggett dredge, known as Leggett No. 3, was put in commission in the spring of 1904, and dismantled in June, 1908, after having worked out the property. The machinery was then taken to Wyman's Ravine, and installed upon a new hull. It was a 5-cubic-foot open-link-bucket elevator dredge, constructed by the Risdon Iron Works.

During the five and one half years this dredge was in operation it handled about 4,500,000 cubic yards of gravel, and turned over 75 acres of ground, of an average depth of about 30 feet. The company employed about thirteen men.

The hull of this dredge was 88 feet long, 30 feet wide and 7 feet deep, with a draught of 4 feet. The digging ladder was solid-girder construction, 58 feet long, carrying 35 buckets, each of 5-cubic-foot capacity, dumping 13 per minute. The revolving screen was 21 feet 6 inches long and 4 feet 6 inches in diameter. The electric motor equipment installed upon the dredge had a rated capacity of 175-horsepower, distributed as follows: Pressure pump, 50-horsepower; primary pump, 50-horsepower; revolving screen motor, 10-horsepower; bucket drive, 75-horsepower; stacker, 10-horsepower; and winch 25-horsepower.

No. 122. Constructing hull for the Leggett Mining Company's Risdon type dredge, at Wyman's Ravine, near Oroville, Butte County.

The average power consumed while in full operation amounted to about 95-horsepower.

The gold-saving device was arranged differently from that on any other dredge in the field, and is one on which Mr. Leggett holds a patent. The general arrangement consists of a riffle system composed of 1½-inch angle-irons placed longitudinally with the flow. The spaces between the angle-irons are filled with natural cobbles, as the dredge digs them. They are not paved by hand, as in the ordinary hydraulic sluices, but are allowed free course to lodge themselves where they fall, and for this purpose openings are arranged in the upper end of the screen to allow certain quantities of medium-sized pebbles to fall through for a short period of time immediately after the clean-ups. The sluice system differs from the side distributing riffle tables generally used, being continuous and somewhat similar to the Holmes system. It is said to facilitate rapid clean-ups and avoids the gathering of amalgam on the tops of the riffles for the reason that the natural paved rock riffles cause the gold to sink deep among the pebbles. While in operation small quantities of quicksilver are sprinkled over the sluices from day to day.

WYMAN'S RAVINE DISTRICT.

The Wyman's Ravine district is located in township 19 north, range 4 east, about 4 miles southeast from Oroville. The dredging lands are located along Wyman's Ravine, which empties into Prairie Slough, a tributary to South Honcut Creek, which flows into the Feather River.

No. 123. Gardella Dredge, a 5¾-cubic-foot Risdon boat, digging in 10-foot ground. Wyman's Ravine District.

There are three companies in the district, the Leggett Gold Mining Company, one dredge; L. & J. Gardella, one dredge; and the Garden Ranch Gold Dredging Company, with one dredge, now abandoned.

The Leggett Gold Mining Company began operations in August, 1909. The officers of this company are as follows: President, J. H. Leggett; vice-president, R. S. Kitrick; secretary, M. A. Wells, Oroville, California; treasurer, H. M. Leggett; manager, J. H. Leggett; superintendent, A. J. McShane. The holdings of this company comprise an area of about 520 acres, located north of Palermo, on Wyman's Ravine, in section 5, township 18 north, range 4 east, and sections 32 and 33, township 19 north, range 4 east. Of the total area about 350 acres are estimated as dredgeable. Of this area about one quarter was previously mined by hand and none of it has been under cultivation. The gravel was prospected by means of drills and averages about 10 feet in depth to bedrock. In character the gravel is a fairly clean, loose wash, carrying few large boulders and very little clay, and in places is overlain by a sandy loam. The gravel is said to yield 15 cents per cubic yard.

The Leggett dredge was built during the summer of 1909 and was put in commission some time in September of that year. The machinery for this dredge was taken from the Leggett No. 3 dredge, formerly operated at Oroville by the Leggett Gold Dredging Company, and mounted on a new and smaller hull, built especially for the Wyman's Ravine property by the Risdon Iron Works.

The Gardella property, located in Wyman's Ravine district, is owned

No. 124.   Gardella Dredge, Risdon type.   Wyman's Ravine District, Butte County.

by L. & J. Gardella, father and son partnership; manager, Laurence Gardella, Oroville, California; dredge superintendent, C. Enclow.

The holdings of the company comprise an area of about 440 acres, located in sections 32, 28, 27, township 19 north, range 4 east, on the Fahey and Humphrey property, about five miles southeast of Oroville. The total dredgeable area is estimated at 175 acres, of which about 20 acres have been dredged. The gravel was prospected by means of drills, and averages about 10 feet in depth to bedrock, and in character is fairly loose, carrying, in places, large boulders and some streaks of clay, and is overlain by a sandy loam. It is said to average 16 cents per cubic yard.

The Gardella dredge was put in commission October 12, 1907. It is an open-link-bucket elevator dredge, constructed by the Risdon Iron Works, the general dimensions and equipment of the dredge being as follows: *Hull.*

Length	74 feet
Width	34 feet
Depth	6 feet
Draught	4 feet 6 inches

Mechanical Equipment.

Capacity of buckets	5 cubic feet
Number of buckets per minute	14
Number of buckets in line	28
Weight of buckets, each	1,100 pounds
Weight of links	420 pounds
Length of digging ladder	54 feet
Weight of digging ladder	30,000 pounds
Length of revolving screen	24 feet
Length of bucket stacker	32 feet
Weight of bucket stacker	14,000 pounds

The type and arrangement of the gold-saving tables are Risdon return sluices, the actual riffle surface being about 350 square feet. The electric motor equipment as installed upon the dredge has a total rated capacity of 165-horsepower, a rated output of 84-horsepower, and a kilowatt output of 63 kilowatts. The motor equipment is distributed as follows:

Pressure pump	50 h.p.
Primary pump	5 h.p.
Shaking screen motor	10 h.p.
Bucket drive motor	50 h.p.
Stacker motor	10 h.p.
Winch motor	15 h.p.
Ladder hoist motor	25 h.p.
Rated capacity	165 h.p.

The concern employs about nine men per month.

The Garden Ranch Dredging Company has an area of about 160 acres, located in sections 22 and 23, township 19 north, range 4 east, on

No. 125.   Garden Ranch placer mining dipper dredge.   Out of commission.   Wyman's Ravine, near Oroville.

the upper end of Wyman's Ravine, and about four miles southeast from Oroville.

The Garden Ranch dredge was a Marion placer-mining, dipper machine, and in construction was similar to the one used by the Oroville

Dredging Company. This machine had a capacity of 30,000 cubic yards per month, and cost, complete, about $25,000. It has been idle for some time, having worked out the property.

HONCUT CREEK DISTRICT.

The Honcut Creek district is located on North Honcut Creek, in township 18 north, range 4 east, and about five miles northeast of Honcut, a station on the Southern Pacific Railroad. There are two dredges operating in the district, one belonging to the Kentucky Ranch

No. 126. Kentucky Ranch Gold Dredging Company's boat, Risdon type, operating on Honcut Creek, Butte County.

Gold Dredging Company, a stock company, and the other to Laurence Gardella.

The Kentucky Ranch Gold Dredging Company began operations May 1, 1909, with one dredge. The company is incorporated under the laws of Arizona and is capitalized for $300,000. The officers are as follows: President, L. Gardella; secretary and treasurer, D. C. Howard, Oroville, California; manager, L. Gardella; superintendent, D. Rodrick.

The holdings of this company comprise an area of about 1,000 acres, located in sections 25, 24, and 19, township 18 north, range 4 east, on North Honcut Creek, about 12 miles southeast of Oroville. The dredgeable area is estimated at about 200 acres. The gravel was prospected by means of drills, and is said to yield about 17 cents per cubic yard, the average depth to bedrock being about 18 feet; the bedrock is a schist.

The Kentucky Ranch dredge is of the open-link-bucket elevator type, constructed by the Risdon Iron Works. The hull of this dredge is 76 feet long, 36 feet wide, 7 feet deep, and draws 4 feet 6 inches. The digging ladder is steel-girder construction, 58 feet long, weighs 36,000 pounds, and carries 32 buckets of 5-cubic-foot capacity, weighing, each, 1,100 pounds, and dumping at the rate of 15 per minute. The links between the buckets weigh about 420 pounds each. The type and arrangement of the gold-saving tables are double return Risdon sluices, having a riffle surface of about 264 square feet. The revolving screen is 21 feet 6 inches long, and about 6 feet in diameter. The tailing stacker is a Risdon bucket elevator, 44 feet long, weighing 20,000 pounds. The electric motor equipment as installed upon the dredge has a total rated motor capacity of about 175-horsepower, and a rated

No. 127. Placer mining plant on Butte Creek, near Centerville. Now abandoned.

motor output of about 78-horsepower, and a kilowatt output of 57 kilowatts. The motor equipment is distributed as follows:

Pressure pump _____ 50 h.p.
Primary pump _____ 5 h.p.
Revolving screen motor_____ 10 h.p.
Bucket drive motor _____ 75 h.p.
Stacker motor _____ 10 h.p.
Winch motor _____ 25 h.p.

The Gardella holdings are located on the Mahle property, almost joining the Kentucky Ranch Gold Dredging Company's holdings on the south. The dredge was put in operation in February, 1910, and is the same dredge which Gardella operated at Oroville.

BUTTE CREEK DISTRICT.

Butte Creek Consolidated Dredging Company.—This company began operations in the spring of 1909, and has an operating plant of one dredge. It is organized under the laws of the State of California, and is capitalized for $2,500.000. The officers are as follows: President, W. T. Bill, Redlands, California; vice-president, Walter L. Krug, Los Angeles; secretary and treasurer, W. H. Isaacs, 516 Fay building, Los Angeles, California; general manager, H. E. Chesebro, Chico, California.

The holdings of the company comprise an area of 1,450 acres, located in sections 24 and 25, township 22 north, range 2 east, and sections 19, 20, 17 and 8, township 22 north, range 3 east, on Butte Creek, near Diamondville, and about 10 miles northeast of Chico, Butte County, California. Out of the total area about 500 acres are considered dredgeable, but in order to acquire this the company was compelled to take over the non-dredgeable land as well. The ground was prospected by means of shafts and drills. The gravel lies on a sandstone bedrock, and ranges in depth from 13 to 28 feet. The entire deposit is difficult to handle with a dredge, owing to the many boulders and uneven bedrock.

Butte Creek Dredge.—The Butte Creek dredge was put in operation May 1, 1909, and during the first two months in operation handled about 250,000 cubic

No. 128. Butte Creek Consolidated Company's dredge, near Diamondville, Butte County.

yards of gravel in turning over about 8 acres of ground. It is an open-link-bucket elevator dredge, equipped with tail sluices and tail scow, and constructed and designed by Ed L. Smith of New York, and equipped with Taylor Iron and Steel Company machinery. The hull is 109 feet long, 40 feet beam, 9 feet deep, with a draught of 4½ feet. It is 48 feet 9 inches over all to top of spud casing. The digging ladder is of steel construction, weighs 40 tons, is 83½ feet long, and carries 39 11-cubic-foot buckets and 39 links. Each bucket and link weigh, together, two tons. The speed of the buckets ranges from 8 to 12 per

No. 129. 11-cubic-foot buckets, Butte Creek Consolidated Dredging Company's dredge, near Diamondville, Butte County. Showing pipe for greasing lower tumbler on the upper side of ladder.

minute. The revolving screen is manganese steel construction, 12 feet long, 6 feet diameter. This screen separates all the gravel up to 10 inches, which is carried out through a sluiceway; rock larger than 10 inches and up to 4 feet is disposed of by a strong steel chute over the side of the dredge, similar to those used on the early Bucyrus dredges. Water for sluices is furnished from a 15-inch pump of a capacity of 1,000 inches, driven by a 100-horsepower motor. The gold-saving device consists of riffle sets about 4 feet square, 4⅞ inches deep, and covered with manganese steel. There are 36 lineal feet of riffles on the main

11—GD

boat and 120 feet on the tail sluice and tail scow. The tail scow or
sluice boat is equipped with an undercurrent and gold-saving tables of
large capacity for the purpose of saving fine gold. The general gold-
saving arrangement was installed for the purpose of saving nugget gold,
and is said by the manager to be very efficient.

The spuds are 28 inches by 48 inches, 50 feet long, and consist of
one wooden and one steel spud. The steel spud weighs 20 tons, manga-

No. 130. Part of main sluice and undercurrent, showing gold-saving tables on tail scow of
the Butte Creek Consolidated Dredging Company's dredge.

nese steel being used in places where the greatest wear comes. The
electric motor equipment has a rated capacity of 375-horsepower, distributed as follows: Bucket-line drive, 150-horsepower; winch motor,
85-horsepower; sluice pump motor, 100-horsepower; primary pump
motor, 25-horsepower; screen motor, 15-horsepower. All motors are
direct-connected.

The dredge has a capacity of 100,000 cubic yards per month. The
company employs an average of about ten men.

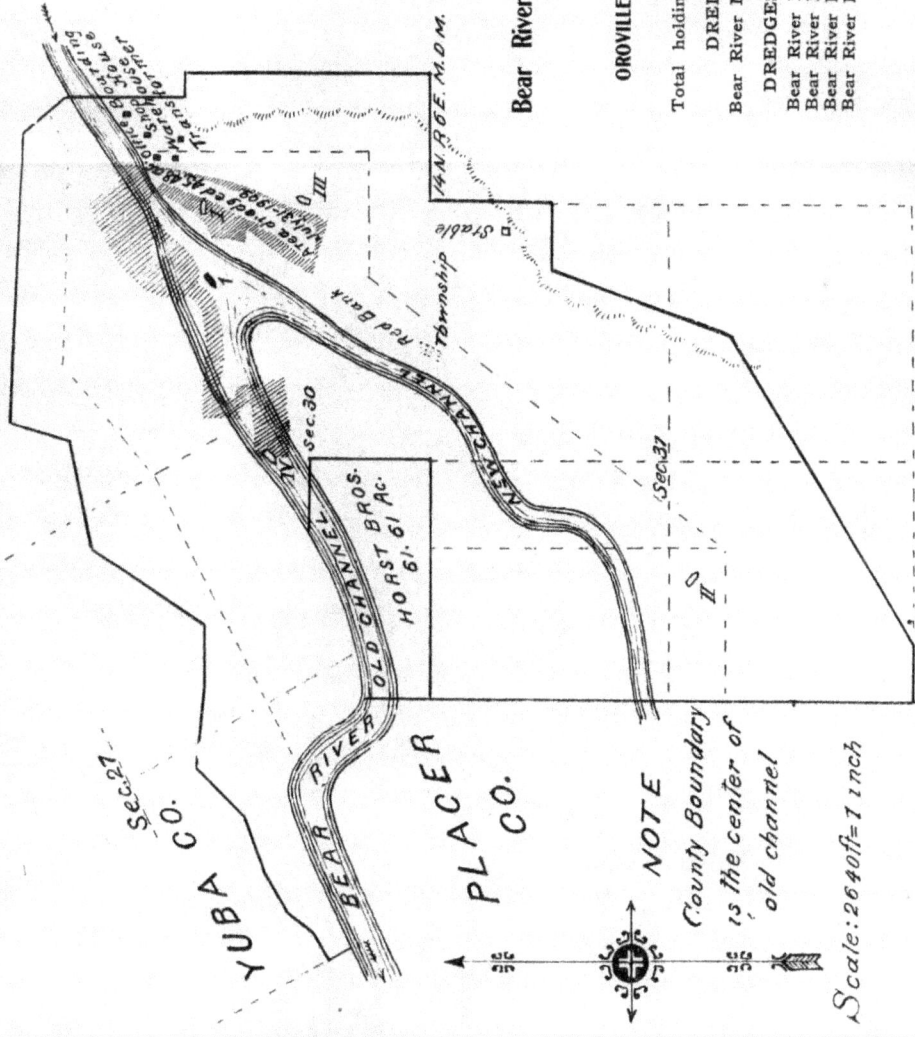

No. 131. Map of Bear River.

2. PLACER COUNTY.

There is only one dredging company operating in Placer County. The property of this company is located in Placer and Yuba counties, near the town of Wheatland, along the Bear River, which drains an area of about 287 square miles between the Yuba and American rivers. The headwaters of the Bear River do not reach back to the crest of the range, so that it seldom receives precipitation under form of lasting snow. It is torrential in character, having no forested areas except in its upper portion. At the head of Bear River ancient gravel deposits are found.

Bear River Mining Company began operations in 1901, and at the present time has an operating plant consisting of one dredge. The holdings of the company comprise an area of 993.10 acres, located in the Bear River Basin, 5 miles from the town of Wheatland, in sections 37 and 30, township 14 north, range 6 east, in Placer County, and in sections 27 and 34, township 14 north, range 5 east, in Yuba County.

Some of this land had been mined previous to dredging, and none of it was ever under cultivation. The property was originally owned by R. D. Evans, of Boston, and W. P. Hammon, of San Francisco, who organized the Bear River Exploration Company, and later the Bear River Mining Company, which took over the holdings of the first company.

Bear River Exploration Company.—Active operations began in 1901, when the Bear River Exploration Company constructed two new hulls, and bought the machinery of two $3\frac{1}{2}$-cubic-foot Risdon steam dredges operating at Breckenridge, Colorado, and originally constructed for the North American Gold Dredging Company in 1897. This machinery was mounted on the new hulls, and electric power installed; but both of these dredges proved too light to handle the heavy ground, and were soon abandoned, after doing little work. Shortly after this, the Bear River Mining Company took over the holdings and installed two new Risdon dredges, especially designed to dig this deposit. However, their operations did not prove profitable, even after changing the bucket-line on No. 2 dredge from 5-cubic-foot open-link, to a $3\frac{1}{2}$-cubic-foot close-connected. Both of these dredges were, therefore, closed down, and work was discontinued until the consolidation with the Oroville Dredging, Limited.

Bear River No. 2.—The dredge now operating was constructed under the present holding company, and is known as Bear River No. 2 dredge. It is a close-connected-bucket elevator dredge of the Yuba Construction Company make, equipped with 7-cubic-foot buckets and heavy machinery. It was commissioned October 1, 1907, and during the first ten months in operation handled 645,095 cubic yards of gravel, and turned over 12.05 acres of ground, at an average cost of 5.41 cents per cubic yard, while digging to an average depth of 33.2 feet.

The property was well prospected by means of drills, many lines of holes being put down at frequent and regular intervals across the channel. The average depth to bedrock is about 40 feet. The gravel averages about 7 cents per cubic yard. It is, in places, very heavy and compact, carrying considerable clay and mud. The bedrock is much the same as that in the Oroville district.

The company employs an average of fourteen men.

3.   YUBA COUNTY.

Yuba County, comprising an area of 625 square miles, or 400,000 acres, with a population of about 13,500 inhabitants, lies in the northern half of the State, and is bounded on the east and south by Sierra and Nevada counties, on the west by Sutter, and on the north by Butte County. Marysville, the county seat, has a population of about 7,000 people, and is located at the junction of the Yuba and Feather rivers, a distance of about 123 miles northeast of San Francisco. It is well supplied with transportation facilities, being on the main line of the Southern Pacific, Western Pacific, and Northern Electric railroads, and it is also well represented by manufacturing establishments, one of the principal of these being the Yuba Construction Company. In the mountainous part of the county mining, lumbering, and stock raising are carried on, while in the valley and foothill portion gold dredging, farming, and fruit raising are the principal industries, the largest hop fields in the West being located at Wheatland. The Yuba River, which passes through the central part of the county, affords an ample supply of water for irrigation projects and power plants.

In 1908 Yuba County had 450 miles of public roads, 44.22 miles of steam railroad, 13.05 miles of electric railroad, one electric power plant, 65 miles of telegraph lines, 2,500 miles of telephone lines, 100 miles of irrigation ditches, and 6,000 acres of land under irrigation.

The following table shows the principal production of Yuba County from 1900 to 1908:

Substances.	1900.	1901.	1902.	1903.	1904.
Brick	------	------	------	------	$3,000
Clay	------	------	------	------	750
Gold	$280,366	$188,908	$155,630	$125,830	139,528
Macadam	------	------	------	------	------
Mineral water	------	------	------	------	------
Silver	4,625	846	2	41	------
Unapportioned	------	------	------	------	------
Totals	$284,631	$189,754	$155,632	$125,871	$143,278

Line of Old State Dam

North Training Wall

No. 12

Levee

CHANNEL OF THE

No. 6

PROPERTY OF Y

Marysville No. 1

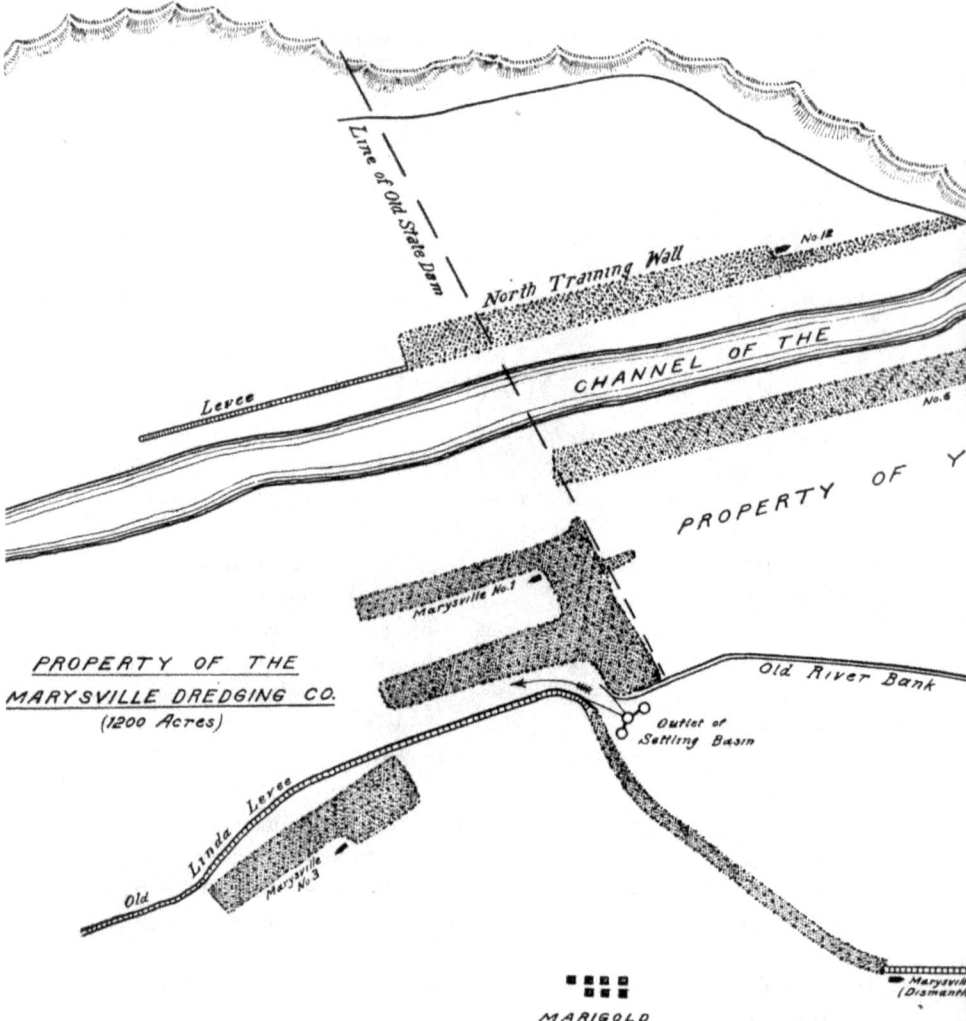

PROPERTY OF THE
MARYSVILLE DREDGING CO.
(1200 Acres)

Old River Bank

Outlet of
Settling Basin

Old Linda Levee

Marysville No. 3

MARIGOLD

Marysville
(Dismantl

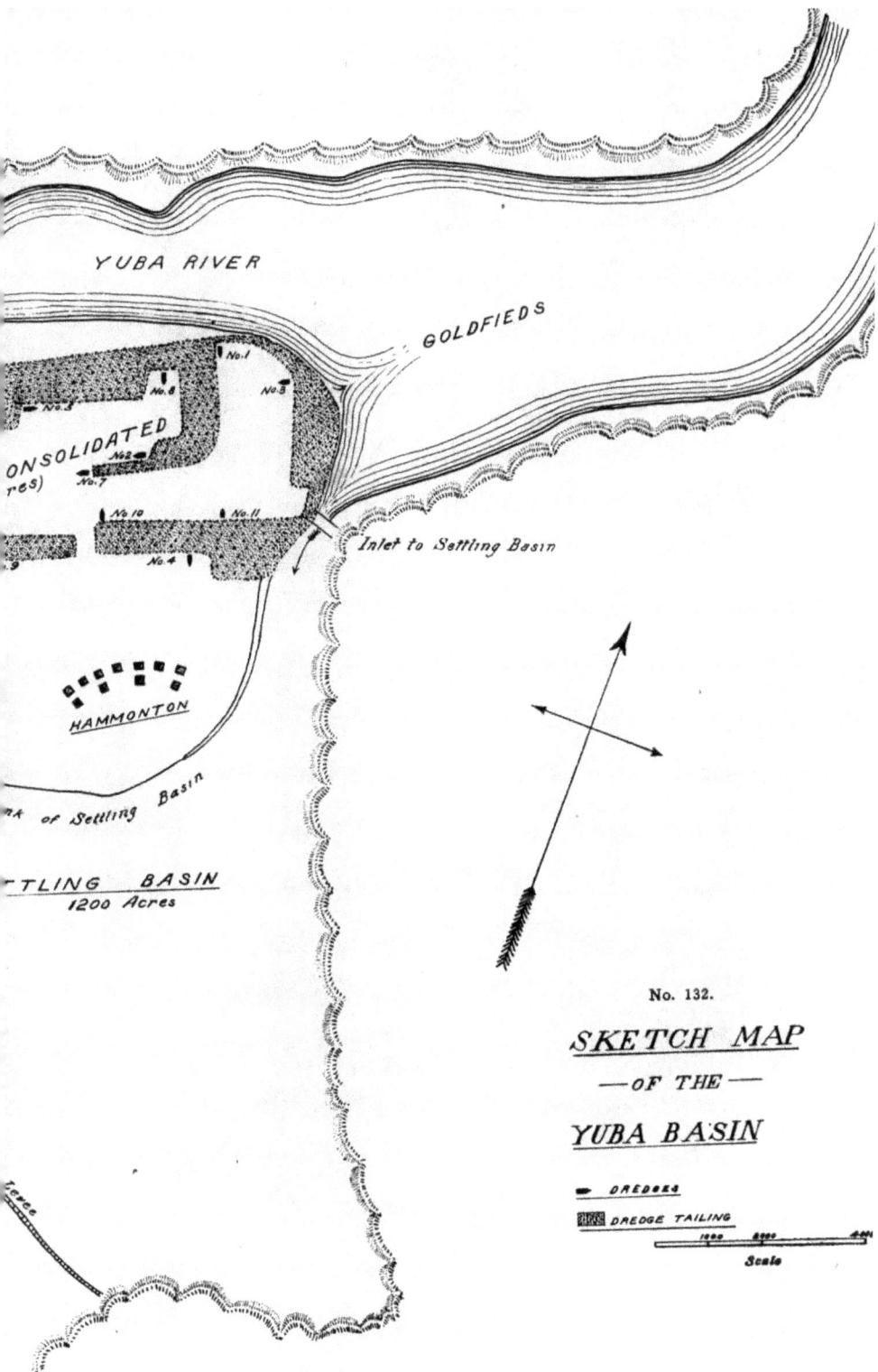

YUBA RIVER

GOLDFIEDS

No.1

No.6

No.3

No.8

ONSOLIDATED
res)

No.2

No.7

No.10

No.11

No.4

Inlet to Settling Basin

HAMMONTON

nk of Settling Basin

TLING BASIN
1200 Acres

No. 132.

SKETCH MAP

— OF THE —

YUBA BASIN

DREDGES

DREDGE TAILING

1000   2000   4000

Scale

Substances.	1905.	1906.	1907.	1908.	Grand Total.
Brick				$10,000	
Clay	$80				
Gold	324,135		$1,766,770	2,034,486	
Macadam				5,750	
Mineral water	800	$800	720		
Silver	369		6,187	9,997	
Unapportioned					$565,004
Totals	$325,384	$800	$1,773,677	$2,060,233	$5,624,304

Since the commencement of gold dredging operations in 1904, at Hammonton and Marigold, on the Yuba River, the gold output of the county has steadily increased until Yuba is now one of the most important gold producing counties in the State.

Gold production from dredging operations in Yuba County from August, 1904, to December 31, 1908:

Year.	Amount.	Increase.	No. Dredges.
1904	$74,263	$74,263	2
1905	188,967	114,704	8
1906	1,205,165	1,016,098	10
1907	1,688,032	482,867	12
1908	1,969,079	281,047	14

In 1904 there was one company operating two dredges during the months of August to December, in 1905 there were eight dredges in operation during part of the year, and in 1906 there were two dredging companies operating part of the year, the Marysville Dredging Company with 2 and the Yuba Consolidated Gold Fields with 8 dredges. During 1907, 12 dredges operated practically the entire year, and during 1908 there were 14 dredges in operation part of the year.

In 1909 the Marysville Dredging Company constructed and put in operation one dredge and dismantled another, so that there were the same number of dredges in active operation as in 1908.

The following table shows the companies, their holdings, and the number of dredges in the field in 1910:

Company.	Dredging Ground, Acres.	Working.	Dredges Dismantled.	Total.
Yuba Consolidated Gold Fields	3,000			12
Marysville Dredging Company	600		1	3
Totals	3,600	1	1	15

Dredging in the Yuba River district began in 1904 when W. P. Hammon of San Francisco and the late R. D. Evans of Boston, Mass.,

No. 133.   Yuba No. 1 and No. 2 dredges leaving construction pit.   Bucket capacity 6 cubic feet.   California type.

commenced operations in August of that year with two California type dredges, the Yuba No. 1 and No. 2.

Previous to the building of these dredges Messrs. Hammon and Evans spent about two years in examining and acquiring the ground located in and adjacent to the Yuba Basin. During this time about 300 test holes were sunk with drills, to depths of from 60 to 70 feet and nearly $100,000 was expended in the work. The ground prospected extended over an area nearly five miles long with an average width of about one mile. The widest part of the property is located about four miles from where the Yuba River leaves a narrow canyon in breaking from the foothills of the Sierra Nevada, and spreads over a wide flat area about two miles across, known as the Yuba Basin.

The depth of the ground is due to the extensive hydraulic mining operations which for years were carried on along the headwaters and tributary streams of Yuba River. It is estimated that aside from the natural sedimentation due to erosion, that nearly half a billion cubic yards of hydraulic tailings have been carried down yearly by the flood waters and deposited in the river valley. The hydraulic tailings in the Yuba Basin range in depth from 10 feet to 40 feet and overlie the old gold-bearing river gravel. These tailings carry some gold, but not enough to pay for working.

The old river gravel underlying the hydraulic tailings rests on a volcanic ash bedrock, similar in character to that of the Oroville district. In some places drill holes showed gravel and some gold below this bedrock, 90 feet and 110 feet from the surface, but usually the volcanic ash lies on the true bedrock.

The gold sometimes forms several pay-streaks in the old river gravel but is seldom found to extend to the volcanic ash bedrock. The ground yields from 10 cents to 30 cents per cubic yard from top to bottom. In dredging part of the bedrock, which is soft and sticky like the false bedrock in other districts, is usually dug. The natural water level ranges from several feet above to about four feet below the surface of the ground.

The Yuba No. 1 and No. 2 dredges which were installed for W. P. Hammon and R. D. Evans were designed and constructed by the Bucyrus and Western Engineering and Construction companies. They were the first dredges of their kind to be built to dig 60 feet below the water-line, and many new features and improvements were brought out in the construction of the hulls and mechanical equipment.

In 1905 the Yuba Consolidated Gold Fields was organized to take over the holdings acquired by W. P. Hammon and R. D. Evans. This company was incorporated under the laws of the State of Maine, with a capitaliza-

No. 134. Yuba No. 7 Dredge, digging 65 feet deep. Bucket capacity 7½ cubic feet. California type.

tion of $12,500,000. The officers are: President, Geo. L. Huntress; secretary and treasurer, R. E. Paine, 50 Congress street, Boston, Mass.; managing director, W. P. Hammon; general manager, Newton Cleaveland, Marysville, Cal.; superintendent Geo. J. Carr, Hammonton, Cal. The holdings of the company comprise an area of 3,000 acres. See map.

All of the 12 dredges operated by the company are of the California type. The Yuba No. 1 and No. 2 dredges have a bucket capacity of 6 cubic feet, and are equipped with shaking screens, while all the others are equipped with 7½-cubic-foot buckets and revolving screens. The tailing stackers on the Yuba dredges are unusually long, owing to the depth of the ground being dredged.

Yuba No. 3 and No. 4 dredges, which were designed by the Boston Machine Shop Company and the Bucyrus Company, were installed early in 1905 and Yuba No. 5 and No. 6 late in 1905. These dredges were constructed by the Yuba Gold Fields, most of the orders for the excavating machinery being placed with the Bucyrus Company. The bucket lines of No. 5 and No. 6 were furnished by the Taylor Iron and Steel Company. Yuba No. 7 and No. 8 dredges were commissioned in 1906, and were equipped originally with Marion machinery. Yuba No. 9 and No. 10 were commissioned in 1907 and Yuba No. 11 and No. 12 in 1908. Yuba No. 13, a 13½-foot dredge, is under construction. These dredges were designed and constructed by the Yuba Construction Company. The following is a general description of one of the latest dredges:

Yuba No. 11 Dredge.—Digging depth below water line, 65 feet.

Size of Hull.—Length over all, 130 feet; width on water line, 46 feet; width of housing, 56 feet; depth, 9½ feet; draught, 6½ to 7 feet. Total lumber in hull, about 375,000 board feet.

Mechanical Equipment.—Length of digging ladder, 118 feet; capacity of buckets, each 7½ feet; weight of buckets, each 2,000 pounds; buckets dumping per minute, 18; total buckets in line, 96; length of tailing stacker, 132 feet; width of stacker belt, 32 inches; revolving screen, width 6 feet, length, 36 feet; spuds, 24 by 36 inches by 55 feet; total weight of machinery, 875,000 pounds; total weight of dredge, 2,000,000 pounds. The spuds are constructed of ⅜-inch side and web plates and ⅞-inch cover plates and angle iron joints; upper and lower tumblers are hexagonal. The digging ladder is truss girder design.

Yuba No. 11 dredge has a rated motor capacity of 405 horsepower and an average motor output of 315 horsepower. Transformer capacity 375 kilowatts and the transformer output 262 kilowatts. The electric current is brought over the primary line to dredge at 4,000 volts and stepped down to 440 volts for all motors. The motor equipment aboard the dredge is distributed as follows:

Bucket drive motor, 200-horsepower; G. E.; 400 volt; 600 revolutions per minute; average output, 137 kilowatts; 148-horsepower. Winch motor, 25-horsepower; G. E.; V. S.; 440 volt; 600 revolutions per

No. 135. Yuba No. 12 Dredge. Yuba Construction Company design. California type.

minute: average output, 13 kilowatts; 17.4-horsepower; while hoisting steel spud, 15.8 kilowatts; 21-horsepower. Revolving screen, 25-horsepower; G. E.; C. S.; 440 volt; 600 revolutions per minute; average output, 16 kilowatts; 24-horsepower. Tailing stacker, 35-horsepower; G. E.; C. S.; 440 volt; 600 revolutions per minute; average output, 16 kilowatts; 24-horsepower.

Pump motors: Ten-inch high pressure pump, 75-horsepower; G. E.; C. S.; 440 volt; 720 revolutions per minute; average output, 55 kilowatts; 73.5-horsepower.

Ten-inch low pressure pump, 35-horsepower; G. E.; C. S.; 440 volt; 600 revolutions per minute; average output, 8 kilowatts; 10.7-horsepower.

Four-inch primary pumps, 10-horsepower; G. E.; C. S.; 440 volt; 1,200 revolutions per minute; average output, 5.2 kilowatts; 7-horsepower.

The electric transmission line from the Bay Counties Power Company meter pole is of No. 4 copper wire 450 feet long and the shore cable to dredge is 3 conductor No. 4 copper 600 feet long.

The dredges on the Yuba River are coöperating in their work with the Débris Commission in building retaining dams

No. 136.   Idler for taking up slack in bucket line used by Yuba Gold Fields.

to control and confine the flow of the Yuba River in the vicinity of Da Guerre Point. This work is being done for the United States and the State of California.

Hammonton, named after W. B. Hammon, is a settlement established by the company on the south side of the Yuba River nearly opposite Da Guerre Point on the north side. It has over 700 inhabitants and many handsome small buildings occupied by the dredging population, and some of the company's offices and repair shops are located here. A well appointed schoolhouse, a general store, etc., water towers for furnishing water for domestic purposes and fire protection, good gravel metaled streets, plenty of shade trees and a regular mail service, etc., all go to make this an attractive little town.

The Marysville Dredging Company began operations in 1905 under the direction of R. E. Cranston, Sacramento, Cal. The officers of the company are: Directors, F. Lothrop, president; Q. A. Shaw, vice-

No. 137. Marysville No. 2 Dredge. Marysville Gold Dredging Company, Yuba River. California type.

president; Alexander Agassiz; T. L. Livermore; H. L. Higginson; R. L. Agassiz, treasurer, 14 Ashburton Place, Boston, Mass.; A. D. Snodgrass, cashier; A. B. Strock, superintendent, Marigold, Cal.; Bukeley Wells, general manager, Marysville, Cal.

The holdings of the company adjoin those of the Yuba Consolidated Gold Fields on the west and comprise an area of about 1,200 acres located in section 36, township 16 north, range 4 east, and sections 1 and 2, township 15 north, range 4 east along the Yuba River, about 6 miles northeast of Marysville. Of the total acreage about 600 acres are dredging ground, averaging in depth about 60 feet. The gravel and bedrock is the same in character as that of the Yuba Consolidated Gold

No. 138. Yuba Construction Company, Marysville, Cal.

1. Machine shops, reinforced concrete, 100 by 250. 2. Forge shop, steel frame, 100 by 100; structural shop, 100 by 150, equipped with all modern tools, specially selected for dredge construction. 3. Powerhouse. 4. Storehouses.

Fields. This company is also coöperating with the Débris Commission in their work.

In 1906 the company installed two California type dredges of 7½-cubic-foot capacity, the Marigold No. 1 and No. 2. These dredges were designed by the engineers of the Yuba Construction Company, and a large portion of the machinery came from the Marion Steam Shovel Company.

In 1909 the company dismantled the Marigold No. 2 dredge and in August of that year installed a new California type dredge, the Marigold No. 3. This dredge was designed and constructed by the company's engineer, L. Ettrup, the machinery orders being given to different manufacturers, principally the Union Iron Works.

On the property of the Marysville Dredging Company a small settlement, known as Marigold, was established by the company. It comprises many buildings, occupied by the dredging population, and company offices and repair shops.

As a direct outgrowth of the interests of the gold dredging industry, there has been established at Marysville, Cal., in the center of the gold

No. 139. Yuba Construction Company, Marysville, Cal. Interior of the forge shop.

dredging fields, a large machine, structural and forge shop, known as the "Yuba Construction Company."

This company was incorporated in 1905, with a capital stock of $200,000, and has the following officers: W. P. Hammon, president; Newton Cleaveland, vice-president and general manager; A. E. Boynton, secretary, Alaska Commercial building, San Francisco; E. P. Jones, superintendent, Marysville, Cal.

Since this shop was organized the investment has been increased from $200,000 to about $500,000. The shop is especially equipped to take care of gold dredging work, and is rapidly becoming one of the largest on the coast. The tools and machinery are of the heaviest construction.

No. 140.   Yuba Construction Company, Marysville, Cal.   Shipping yard, connection with Southern Pacific, Western Pacific, and Northern Electric Railways.

This company has designed and built nineteen of the most successful dredges operating in California, and at the present time is building dredges, not only in California, but in Alaska and several of the Western states.

The accompanying illustrations will give the reader some idea of the extent of the shops, and these are situated on the main lines of the Southern Pacific, Western Pacific, and the Northern Electric railroads.

4.   SACRAMENTO COUNTY.

Sacramento County, comprising an area of 987.66 square miles or 632,108 acres, is one of the largest counties in the Sacramento Valley. It was organized by the first legislature, and holds within its boundaries Sacramento City, the capital of California and the county seat of Sacramento County, situated on the east bank of the Sacramento River, and 90 miles by rail from San Francisco. In 1909 Sacramento City had 55,000 inhabitants out of a total population of 70,000 for the entire county. The total assessed valuation of all property amounted

No. 141. Yuba Construction Company, Marysville, Cal. Interior of structural shop.

to $57,679,076. The total number of acres under cultivation amounted to about 370,000 acres, leaving 262,108 acres not under cultivation. There are no mountains in Sacramento County, the ground rising gradually from an altitude of 30 feet to the low rolling foothills of the Sierra Nevada Mountains, which, at the extreme eastern part of the county, attain an altitude of about 580 feet above sea level.

In 1908 the production of gold from the various mines in Sacramento County amounted to $1,166,055. Of this $409 came from hydraulic mines, $52,365 from drift mines, $3,925 from surface placers, and $1,109,196 from dredging operations.

No. 142. American River from Fair Oaks bridge, Sacramento County.

The following table shows the number of dredges and dredging companies operating in Sacramento County, as well as the number of dredges put in operation or dismantled during 1908:

Company.	Working.	Put in Commission During Year.	Dismantled During Year.
Ashburton Mining Co.	1	1	--
Folsom Development Co.	6	1	1
El Dorado Gold Dredging Co.	1	--	--
Natomas Development Co.	3	3	--
Totals	11	5	1

The total known dredgeable area in Sacramento County in 1908, including that which has been dredged, comprised 6,052.66 acres. The following table gives a comparison of the area and character of the dredgeable and other lands in Sacramento County in 1909:

Character of Lands.	Other Lands, Acres.	Dredging Lands, Acres.
Under cultivation	370,000	2,365
Not cultivated	262,108	3,687.66
Totals	632,108	6,052.66

All dredge mining in Sacramento County is carried on along the American River in what is known as the American River or Folsom district. This district includes in its boundaries a part of the town of Folsom, and extends 9 miles from Folsom to Cornell, a place on the Sacramento and Placerville Railroad, 11 miles by rail from Sacramento City.

The total area of the district comprises 12,522.58 acres, of which about 6,052.66 acres have so far been proven dredgeable. All of this land, with the exception of 1,326 acres on the Sacramento, Sailor, and Mississippi bars, is located on the south side of the American River, extending in places 2½ miles inland. All of the land in the district, with the exception of 431 acres, is owned by the Natomas Consolidated of California, the larger of the two companies now operating in the field.

The production of gold from dredging operations in the American River district from 1899 to December 31, 1908, amounted to $3,920,231, as shown by the following table:

Year.	Gross Value.	Increase.	Decrease	Year.	Gross Value.	Increase.	Decrease.
1899				1904	$348,990	$246,893	
1900	$17,200			1905	569,124	220,134	
1901	47,619	$30,419		1906	921,300	352,176	
1902	155,194	107,575		1907	649,511		$271,789
1903	102,097		$53,097	1908	1,109,196	459,685	

The decrease in 1903, as compared with 1902, was partly due to two dredges being put out of commission—the Ashburton No. 1, which was destroyed by fire while operating on Sailor Bar on May 25; and Pacific No. 1, which was closed down during the year while operating on Mississippi Bar.

The decrease in yield in 1907, as compared with 1906, was partly due to three dredges going out of commission and to severe storms and floods in 1907 which damaged and delayed operations of others.

The increase in yield in 1908 was mainly due to four new dredges being put in commission during that year. They were the Natoma No. 1, No. 2, No. 3, and Folsom No. 6, all large modern boats.

Table Showing Numerical Strength of Gold Dredges in Sacramento County at the End of the Year 1909.

Company.	Working.	Idle.	Constructing.	Total.
Natomas Consolidated of California	7	1	1	9
Ashburton Mining Company	1	--	--	1
Totals	8	1	1	10

AMERICAN RIVER DISTRICT FROM 1899 TO 1909.

Colorado Pacific Gold Dredging Company.—Dredging in the American River district began in the spring of 1899, when the Colorado Pacific Gold Dredging Company, promoted by R. G. Hanford, and consisting principally of Colorado mining men, commenced operations in April of that year with one dredge on the company's holdings on the north side of the river, on what is known as the Mississippi Bar.

Pacific No. 1 was an open-link-bucket elevator dredge, built by the Risdon Iron Works, for a capacity of 35,000 cubic yards per month, and equipped with 3½-cubic-foot buckets. It was the only steam driven dredge in the district, and was kept in operation for about four years until dismantled in 1903. See page 180.

Pacific No. 2.—In 1902, a second dredge, known as Pacific No. 2, was put in operation on the property. This dredge was of the same make as No. 1, but had 5-cubic-foot buckets, and was electrically driven, and had a rated capacity of 60,000 cubic yards per month. It was dismantled in 1906 after about four years of service.

Ashburton Mining Company.—The second company to commence dredging operations in the district was the Ashburton Mining Company, also promoted by R. G. Hanford. Under the direction of R. E. Cranston this company began operations March 1, 1900, with one dredge, on the north side of the river, on what is known as Sailor Bar. See page 6.

Ashburton No. 1.—This dredge was of the double-lift-open-link-bucket elevator type, constructed by the Bucyrus Company, and equipped with tail scow and tail sluices 7½-cubic-foot buckets. It worked about four years until destroyed by fire May 25, 1903, when it was reconstructed and operated until dismantled in 1906, after working out the property. See page 179.

Ashburton No. 2.—In 1908 the company constructed a dredge of the single-lift-close-connected-bucket type and installed it on the company's holdings on the south side of the river. This boat, known as Ashburton No. 2, is still in operation, and will probably complete the turning over of the company's holdings of 205 acres.

The gravel on this property ranges in depth from 12 to 48 feet, the average being about 27 feet. In places there are many boulders and a great deal of clay, but the pay streak is usually free from either.

Syndicate Dredging Company.—The third company to begin operations in the district was the Syndicate Dredging Company, which in 1901 put the Syndicate dredge in commission. This dredge, constructed by the Risdon Iron Works, was of the open-link-bucket elevator type, electrically driven, and equipped with 5-cublic-foot buckets. It operated at Natoma, near the middle of the district, where the Natoma No. 1 is now working until closed down in the spring of 1906. R. G. Hanford was the chief stockholder in this company. See p. 182.

Folsom Development Company.— The fourth company to begin operations in the district was the Folsom Development Company, which in 1904 started work with two dredges, the Folsom No. 1 and No. 2. Like the preceding companies, it was launched by R. G. Hanford, the personnel consisting mainly of the Armours and other Chicago people. The company constructed and put in commission a total of six dredges, some of which were the largest of their kind at the time of construction, and also built a large machine shop and erected the first successful rock-

No. 144. Old type Bucyrus double-lift open-link bucket dredge, showing tail scow and tail sluices and rock chute on side of housing. Ashburton No. 1 Dredge, now dismantled.

No. 145. Colorado Pacific No. 1 Dredge, now dismantled. The only steam-driven dredge in the Folsom district. See page 178.

crushing plant to crush the dredge tailings. It was the most successful and progressive company in the district, and was absorbed by the Natomas Consolidated at the end of 1908.

Folsom No. 1 was put in commission February 20, 1904, in the vicinity of Rebel Hill, where it turned over 45 acres of ground of an average depth of 45 feet and handled 2,995,000 cubic yards of gravel, until dismantled August 20, 1908, after four years and six and a half months in operation.

Folsom No. 2 was put in commission March 16, 1904. This dredge worked on the upper end of the district, from Dredge to Nigger Bar, where digging to an average depth of 17 feet, it turned over 150 acres and handled 5,350,000 cubic yards of gravel. It was wrecked by the floods of January 14, 1909, after which part of the machinery was installed on a dredge in Colorado.

Both of these dredges were of the close-connected-bucket elevator type, with 5½-cubic-foot buckets, and were constructed by the Western Engineering and Construction Company, and equipped with Bucyrus machinery. See pages 183 and 184.

These boats, as well as those of the Colorado Pacific Company, were shut down owing to their small capacity and to great improvements in dredge construction.

Folsom No. 3 was put in commission January 1, 1905, and during the first four years in operation turned over about 81½ acres and handled

No. 146. Road cutting near Folsom, showing layer of volcanic ash and gravel. See page 188.

3,600,000 cubic yards of gravel, while digging to an average depth of about 30 feet. It is of the close-connected-bucket type, and was originally equipped with 80 7-cubic-foot buckets, which were later changed to 8½-cubic-foot. At the time of construction this was the largest gold dredge known, and marked the first step toward the large dredge of to-day. Most of the machinery was designed and a great deal of it constructed by the Folsom Development Company, who built the hull and installed the machinery themselves.

This dredge was put in commission on Willow Hill, where it operated intermittently, and not altogether satisfactorily, for a little over two years, when it was turned over to the Bucyrus and Western Engineering and Construction companies for reconstruction. It closed down in May and was recommissioned July 28, 1908, after practically all the machinery had been replaced and new parts installed. Notably among the changes were the new winches that were installed in place of

No. 147. Syndicate Mining Company's dredge at Folsom. Old Risdon type, now dismantled. See page 178.

the old ones, which had given a great deal of trouble and cause for delays. A 13-foot section was added to the digging ladder, which increased the digging depth about 10 feet, and the number of buckets from 80 to 87. The old bucket-line was discarded and replaced by 87 new 8½-cubic-foot buckets. The shaking screens and drive were taken out and a new revolving screen 36 feet long by 7 feet diameter, and new screen drive and screen casing added. A 5-inch two-step hopper

No. 148. Arrangement to prevent thefts from principal part of gold-saving tables. Folsom District.

pump, with piping to furnish five sprays in the hopper, and two 8-inch two-step monitor pumps, with motors and piping, as well as a new save-all in the well-hole, were also installed.

No. 149. Folsom No. 2 Dredge in operation. Early Bucyrus design. Dismantled.
See page 180.

The increased running time and general efficiency of Folsom No. 3 dredge after reconstruction brought the yardage up from 70,000 cubic yards to 180,000 cubic yards per month. This dredge is in good working condition after five years in operation up to January 1, 1910.

No. 150. Wreck of Folsom No. 2 Dredge. See pages 180 and 183.

The electric motor equipment of the No. 3 dredge is complete and mechanically in good condition. It has a rated capacity of 490-horse-power, distributed as follows:

Bucket drive motor	150 h.p.	V.S.	345 r.p.m.	W.H.	2000 volts
14-inch pump motor	100 h.p.	C.S.	580 r.p.m.	W.H.	2000 volts
7- and 8-inch pump motors	50 h.p.	C.S.	850 r.p.m.	W.H.	2000 volts
Monitor pump motors, two 50 h.p.	100 h.p.	C.S.	850 r.p.m.	W.H.	2000 volts
Screen motor	30 h.p.	C.S.	850 r.p.m.	W.H.	400 volts
Winch motor	30 h.p.	V.S.	850 r.p.m.	W.H.	400 volts
5-inch step hopper pump motor	30 h.p.	C.S.	850 r.p.m.	W.H.	400 volts
Rated capacity	490 h.p.	Folsom No. 3 dredge.			

No. 151. Folsom No. 3 Dredge, showing 8½-cubic-foot close-connected buckets and ladder construction.

The average power consumed while in full operation is 401.5 horse-power, or 299.4 kilowatt. The monitor pumps are not in use.

The close relation of the individual parts to the efficiency of the whole is clearly shown in the reconstruction of this dredge.

Folsom No. 4 began operations November 15, 1905. This dredge is of the close-connected-bucket elevator type, and equipped with 68

No. 152. Splicing additional section in digging ladder of Folsom No. 3 Dredge to increase digging depth. See page 182.

13-cubic-foot buckets. It is working near the west end of the district on low beach land, and up to January 1, 1909, or during a little over three years, has turned over 209 acres and handled 6,500,000 cubic yards of gravel, while digging to an average depth of 20 feet.

Folsom No. 4 was the first gold dredge of its size ever constructed, as well as the first dredge to be equipped with double-bank gold-saving tables. The general idea of the construction originated with R. G. Hanford, then general manager of the Folsom Development Company, who, contrary to the advice of experienced dredge men and dredge constructors, carried out his ideas for the construction of this dredge.

No. 153.   Folsom No. 4 Dredge, showing bow gantry and bucket-line, 13-cubic-foot buckets.

No. 154.   Winch House, Folsom No. 4 Dredge, Folsom District.

As the washing surface of a dredge is limited to the size of the hull, it became a serious factor how to wash the large amount of material due to the increased capacity of the buckets. This difficulty was overcome by the arrangement of the double-bank tables, which were worked out by Hanford's dredge superintendent, S. A. Martindale, and installed on the dredge. Aside from many important changes the screening area was increased by the addition of large shaking screens, and all parts of the dredge strengthened and made heavier in proportion to the increased capacity of the buckets. The dredge was partly designed and constructed by the company at the Folsom machine shops, Bucy-

No. 155. Folsom No. 5 Dredge, showing double tail sluices, California type.

rus machinery being installed. It has handled about 250,000 cubic yards per month, at a cost of about 3 cents per cubic yard, and is in good working condition to-day after over four years in operation.

The electric motor equipment installed upon Folsom No. 4 dredge has a rated capacity of 415-horsepower, distributed as follows:

Bucket drive motor	200 h.p.	580 r.p.m.	W.H.	V.S.	2000 volts
16-inch Worthington pump motor	100 h.p.	580 r.p.m.	W.H.	C.S.	2000 volts
Winch motor	30 h.p.	850 r.p.m.	W.H.	V.S.	400 volts
Shaker motor	30 h.p.	850 r.p.m.	W.H.	C.S.	400 volts
Stacker motor	40 h.p.	850 r.p.m.	W.H.	V.S.	400 volts
5-inch primary pump motor	15 h.p.	1120 r.p.m.	W.H.	C.S.	400 volts
Rated capacity	415 h.p.				

The average power consumed while in full operation is 260.2 horsepower, or 195 kilowatts.

Folsom No. 5 was put in operation December 10, 1905. This dredge is of the close-connected-bucket elevator type, equipped with 73.9-cubic-foot buckets, monitors, and double-bank tables. It is working on Rebel Hill, digging in partly cemented gravel to an average depth of about 60 feet, and up to January 1, 1909, or a little over three years, has turned over 39 acres and handled 3,350,000 cubic yards of gravel.

The gravel in the Rebel Hill section is deeper and more compact than any other in the district. It ranges in depth from 50 to 75 feet, is partly cemented for the first 6 or 8 feet, and compact for the next 25 or 30 feet, but the lower portion is loose and does not present much difficulty in dredging.

To dredge this deposit without constant blasting was long considered economically impossible, even at a depth of 30 feet, and to construct a dredge with 9-cubic-foot buckets to dig to a depth of 70 feet in a deposit of this kind, with all the excessive weight of machinery that would follow such a design, was thought impracticable. The problems

No. 156. Folsom No. 5 Dredge, showing lower tumbler and bucket line, also monitor on bow of dredge. Folsom District.

presented were overcome by R. G. Hanford, who decided to use monitors to break down the first 20 or 30 feet of gravel, thus leaving only the looser gravel to be dug. The Western Engineering and Construction Company constructed the dredge and installed Bucyrus machinery.

Folsom No. 5 was the first gold dredge to be designed with monitor equipment. It has two monitors with three-inch nozzles placed on the bow, one on each side of the well-hole, and supplied with water from a high-head centrifugal pump driven by two 50-horsepower motors. Owing to the water-level of the pit being sometimes 20 feet below the surface of the ground, and the first six or eight feet of top gravel cemented, it was found difficult to dredge this ground without injury to the dredge and machinery, especially the ladder, as careless handling results in large blocks of cemented gravel falling upon the ladder and bucket-line. While the monitors proved successful in under-mining and breaking down the cemented gravel above the water-line and the digging capacity of the boat was sufficient to take care of the material, it soon became apparent that the washing facilities were inade-quate. The Western Engineering and Construction Company under-took to reconstruct the dredge in part and made several important changes.

In reconstructing the shaking screens were removed and new revolv-ing screens and drives installed, one upper bank of tables and one new

No. 157. Steel casting spud guide to protect stern on Folsom No. 5 Dredge.

longitudinal sluice on each side of the boat were added, and a new device put in the save-all in the well-hole. Repairs were made to ladder, main drive, sheaves, spuds, and hull, etc. The 150-horsepower digging motor was replaced by a 200-horsepower motor, and, as on Folsom No. 3, a heavy steel casting was placed on the stern of this dredge for the purpose of taking up the wear and tear caused by the steel spud rubbing against the wooden hull.

The general equipment of the dredge to-day is as follows: Seventy-three 9-cubic-foot buckets, weighing 2,853 pounds each, and dumping 18 per minute; plate-girder ladder, weighing 132,980 pounds; double-bank gold-saving tables, upper set wood and lower steel construction; revolving screen 7 feet diameter by 36 feet long; steel lattice stacker frame, 138 feet long, weighing 39,280 pounds; conveyor belt 40 inches wide, 285 feet long.

The hull is 110 feet long, 44 feet 7 inches wide, and 10 feet deep, and was especially designed to stand the heaviest digging strains. The contract digging depth is 35 feet below water-line.

No. 158. Showing upper and lower tail sluices and long tailing conveyor, Folsom No. 5 Dredge.

The electric motor equipment installed on Folsom No. 5 has a rated capacity of 540 horsepower, distributed as follows:

Bucket drive motor	200 h.p.	V.S.	W.H.	580 r.p.m.	2000 volts
14-inch pump motor	100 h.p.	C.S.	W.H.	580 r.p.m.	2000 volts
Monitor motors, two 50 h.p.	100 h.p.	C.S.	W.H.	850 r.p.m.	2000 volts
Winch motor	30 h.p.	V.S.	W.H.	850 r.p.m.	400 volts
Screen motor	30 h.p.	C.S.	W.H.	850 r.p.m.	400 volts
Stacker motor	30 h.p.	C.S.	W.H.	850 r.p.m.	400 volts
Two-step hopper pump motor	30 h.p.	C.S.	G.E.	900 r.p.m.	440 volts
4-inch priming pump motor	20 h.p.	C.S.	W.H.	1100 r.p.m.	400 volts
Rated capacity	540 h.p.				

The average power consumed while in full operation is 484.8 horsepower, or 363.8 kilowatts. The 4-inch priming pump is seldom used.

After the double-bank tables were worked out on the dredge and the long upper tail sluices installed, it was found that the large sand pumps were not necessary to take care of the tailing, and they are, therefore, not in commission.

Folsom No. 6.—The sixth and last dredge to be put in operation by the Folsom Development Company was Folsom No. 6. This dredge began operations March 8, 1908, on Sulkey Flat, near Rebel Hill, where it is still digging, in partly cemented gravel, to an average depth of 60 feet, and against a bank 20 feet high above the water-line. During the year ending December 31, 1909, it turned over 18.9 acres and handled 1,565,598 cubic yards of gravel at a cost of 5.8 cents per cubic yard.

This dredge was especially equipped to dig tight and partly cemented gravel, and no expense was spared to make it as effective as possible. The hull was designed to withstand the heaviest digging strains and has proven one of the staunchest of its kind.

No. 159. Folsom No. 6 Dredge digging 75 feet, showing long tailing conveyor and heavy ladder construction. Bucyrus and Western Engineering and Construction Company, builders. California type.

A general description of the hull and equipment of this dredge is as follows: Digging depth below water-level, at 45 degrees 48 feet.

Hull.

Length	120 feet
Width on water-line	46 feet 6 inches
Depth	10 feet 3 inches
Draught	6 feet 10 inches

Mechanical Equipment.

Capacity of buckets	9 cubic feet
Number of buckets in line	86
Weight of buckets	262,816 pounds
Weight of buckets, each	3,056 pounds
Weight of upper tumbler	20,000 pounds
Weight of lower tumbler	16,000 pounds
Weight of ladder and fittings	144,000 pounds
Weight of revolving screen	46,410 pounds
Dimensions of revolving screen	7 feet diameter, 36 feet long
Length of stacker	142 feet centers
Stacker belt	38 inches wide, 293 feet long
Weight of steel spud	76,000 pounds
Dimensions of steel spud	34 by 54 inches by 75 feet long
Dimensions of wooden spud	34 by 54 inches by 78 feet long

The wooden spud is one of the largest single sticks of timber ever handled on the Pacific coast. The digging ladder is of the plate-girder type, and the tailing stacker lattice-girder construction equipped with Robins belt conveyor. See pages 55 and 60.

In addition to the machinery weight the following is a list of some of the material used in construction of the dredge:

Iron and steel	30,000 pounds
Truss rods and plates	30,000 pounds
Piping and valves	24,000 pounds
Nuts, bolts, and nails	33,000 pounds

The weight of machinery on the dredge is 1,253,700 pounds, and the total weight of lumber in hull is 1,141,535 pounds, making the total weight of the dredge 2,395,235 pounds.

The electric motor equipment installed upon the dredge has a rated capacity of 790-horsepower, distributed as follows:

Main digging, or bucket drive motor_____200 h.p.   V.S.   600 r.p.m.
Main pump, 12-inch high-pressure motor_____100 h.p.   C.S.   580 r.p.m.
Secondary pump, 12-inch low-pressure motor_____ 50 h.p.   C.S.   580 r.p.m.
12-inch Monitor pump motors, two 150 h.p._____300 h.p.   C.S.   580 r.p.m.
Revolving screen motor_____ 30 h.p.   V.S.   850 r.p.m.
Winch motor _____ 30 h.p.   V.S.   850 r.p.m.
Stacker motor _____ 50 h.p.   V.S.   850 r.p.m.
5-inch two-step hopper pump motor_____ 30 h.p.   C.S.   850 r.p.m.

   Rated capacity _____790 h.p.

The average power consumed while in full operation is 572-horsepower, or 428.3 kilowatts.   All motors 50-horsepower and over are

No. 160.   Steel spud, 34 inches by 54 inches by 75 feet long.   Folsom No. 6 Dredge.

Westinghouse 2,000-volt, and all motors under 50-horsepower are Westinghouse 400-volt.

, A more detailed description of the principal pumps is as follows: Main or high pressure pump, one 12-inch 60-foot head centrifugal, direct-connected to one 100-horsepower motor; secondary or low pressure pump, one 12-inch 30-foot head centrifugal, direct-connected to 50-horsepower motor.   Pump for monitors, 150-foot head centrifugal, direct-connected, and driven by two 150-horsepower motors.

The two monitors are operated either singly or together.   Care and judgment must be exercised in washing down the bank, for should too much gravel be loosened at one time, the "cave" might fall too far behind the bow of the dredge for the buckets to pick up the gravel, or else it might be necessary to handle too great a quantity of material at one setting, and thus ground the dredge at the stern.

The double-bank tables, which are of steel construction, afford ample room for gold-saving purposes and do away with sand troubles and the need of sand pumps. The tail sluices are 4 feet wide. See pages 73 and 74.

No. 161. Framing large wood spud for Folsom No. 6 Dredge, size 34 by 54 inches, 75 feet long, weighing 35,000 pounds.

The three-eye type buckets have not proven satisfactory. They cause great wear on the upper-tumbler faces, and are, therefore, a source of more or less annoyance and delay; the main cause for delay in the operation of this dredge, however, comes from the bow-swing lines,

No. 162. Placing point on large wood spud. Folsom No. 6 Dredge.

which at the outset were ⅞ of an inch in diameter, but owing to the hard usage and short life were replaced by 1⅛-inch lines, which, though cumbersome to shift, are giving fair satisfaction. The side-line trouble is increased on account of the high bank which is always around the

13—GD

dredge. The shore-sheave hangs over the side of the bank on a pendant, but it is not practicable to get a sheave large enough for the line because of the up-and-down lead to the line which chafes on the flange as well as on the rim of the sheave. The weight of such a sheave would increase the difficulty of shifting the lines.

No. 163. Two-step hydraulic monitor pumps, direct-connected to Westinghouse motors. Folsom No. 6 Dredge.

Recently the complete line of three-eye buckets was replaced by a new line of buckets made with single-eye in rear and two forward eyes, similar in design to the bucket used on Folsom No. 5 dredge. See illustration No. 156, page 188.

No. 164. Hydraulic jets breaking down bank in front of Folsom No. 6 Dredge.

The construction of Folsom No. 6 followed the successful operations of Folsom No. 5, which led R. G. Hanford to contract for the building of a still larger and heavier dredge to work the partly cemented gravel deposit of Rebel Hill. The plans for Folsom No. 6 were worked out by the Western Engineering and Construction Company, who constructed the dredge and installed Bucyrus machinery.

The development of Folsom No. 5 and No. 6 introduced many new factors in dredge construction, and marked a distinct forward step in gold dredging, in so much as they demonstrated that ground which a few years ago was considered impossible to dredge, and could not then have been dredged economically, can now be worked at a profit. The success of these dredges materially increased the boundaries of the district, as up to that time dredging had been confined to the lower bench lands where the gravel is comparatively loose, and free from clay.

To give credit for the success of a number of things with one man is often difficult and not always just towards others who assisted in the perfecting of the details which are necessary for the success of the whole. However, in this instance, as the moving spirit of the whole, the credit for the success of gold dredging and the improvements in dredge construction in the American River district up to 1909 must be

No. 165.   Digging ladder before assembly.   Folsom No. 6 Dredge.

given to R. G. Hanford, who since 1899 has promoted and managed the affairs of all but one dredging company in the district.

El Dorado Gold Dredging Company.—The fifth company to start operations in the district was the El Dorado Gold Dredging Company, the personnel of which consisted of San Francisco men, with E. H. Benjamin secretary. This company began operations April 25, 1905, with the El Dorado dredge, a Risdon built boat of the open-link-bucket elevator type, equipped with 7-cubic-foot buckets. The company's holdings consisted of 554 acres on the south side of the American River, at the west end of the district. The dredge worked intermittently on this property until some time in 1908. In the spring of 1909, it was taken over by the Natomas Consolidated of California, who operated the dredge until May 31st of that year, when it was again shut down and remained idle practically the rest of the year. This dredge is now permanently out of commission.

Natoma Development Company.—The sixth company to start operations in the Folsom district was the Natoma Development Company, which took over the holdings of the Syndicate Dredging Company, the Colorado Pacific Gold Dredging Company, and the Natomas Vineyard Company.

During 1908 this company commissioned three dredges, Natoma No. 1, No. 2 and No. 3, all close-connected-bucket elevator dredges, designed and constructed by the Yuba Construction Company, who ordered much of the excavating machinery for these dredges from the Bucyrus Company.

Natoma No. 1 dredge, the largest gold dredge operating in California at the present time, was put in commission May 10, 1908, on the Syndicate property in the pit of the old Syndicate dredge. It required 420,000 feet of lumber in the construction of the hull, which is 112 feet long, 50 feet wide on water line, 11 feet 6 inches deep, with a draught of 8 feet 6 inches. The hull is strengthened by two lateral and two fore and aft trusses running the full length of the boat. The digging

No. 167. Natoma No. 1 Dredge, showing 13½-cubic-foot buckets and spuds. California type.

depth is 35 feet below water line. The digging ladder is plate girder type, 104 feet long, and carries sixty-one 13½-cubic-foot buckets weighing each 3,450 pounds and dumping 20 per minute. The tumblers are hexagonal. The upper shaking screen is 10 feet 10 inches wide by 22 feet long, and the lower 11 feet 8 inches wide by 22 feet long. The gold-saving tables are double bank, constructed of wood and equipped with wood riffles of the Hungarian type, the total riffle area being 5,500 square feet. The tailing stacker is 104 feet long and carries a 42-inch belt. Each of the two steel spuds is 36 inches by 52 inches by 55 feet long and weigh each 68,000 pounds.

Electric Motor Equipment Natoma No. 1 Dredge.

	Rated Capacity.	Type.		Average Output.		
Main drive motor	300 h.p.	V.S.	514 r.p.m.	150 k.w.	200 h.p.
Winch motor	35 h.p.	V.S.	600 r.p.m.	21.4 k.w.	28.5 h.p.
Shaker motor	75 h.p.	C.S.	720 r.p.m.	30 k.w.	40 h.p.
Stacker motor	35 h.p.	C.S.	600 r.p.m.	18.6 k.w.	25 h.p.
14-inch high-pressure pump motor	150 h.p.	C.S.	600 r.p.m.	73.2 k.w.	98 h.p.
10-inch low-pressure pump motor	35 h.p.	C.S.	600 r.p.m.	26.2 k.w.	35 h.p.
5-inch priming pump motor	15 h.p.	C.S.	1200 r.p.m.	14 k.w.	18.7 h.p.
Totals	645 h.p.			333.4 k.w.	445.2 h.p.

All motors are General Electric Company 550-volt.

The hull of Natoma No. 1 dredge was built to accommodate extra large shaking screens and double-bank gold-saving tables, which were installed to insure a clean washing and distribution of the great amount of material handled by the large buckets. As the gravel is clean and carries but a small amount of sand, it washes easily and runs freely over the riffles.

During the first year in operation, in the months of May, June, and July, respectively, this dredge handled an average of 8,029, 8,200, and 8,350 cubic yards of gravel per day of twenty-four hours, at a cost of 2¼ cents per cubic yard. The power used averaged 39,200 kilowatt hours per week, or about 3-5 of a cent per cubic yard.

During the first eight months in operation this dredge turned over 60.02 acres of ground and handled 1,830,000 cubic yards, during 235 operating days, at a cost of 2.3 cents per cubic yard, while digging to an average depth of 19 feet. During the twelve months ending December 31, 1909, the dredge handled 3,048,-254 cubic yards of gravel, or an average of 8,397 cubic yards

No. 168. Belt conveyor for stacking dredge tailing. Natomas No. 2 Dredge.

per day, at a cost of 2.41 cents per cubic yard, while digging to an average depth of 19.07 feet.

Natoma No. 2 dredge began operations April 22, 1908. It is working on the north side of the river on what is known as Sacramento Bar. It is a close-connected-bucket elevator dredge, equipped with 8-cubic-foot buckets.

During the first eight months in operation this dredge turned over 16.38 acres, handling 626,300 cubic yards of gravel during 130 operating days, at a cost of 2.4 cents per cubic yard, while digging to an average depth of 24 feet in fairly loose gravel. During 12 months in 1909 the dredge operated 6,849.05 hours, handling 1,962,448 cubic yards, an average of 5,406 cubic yards of gravel per day, at a cost of 3.1 cents per cubic yard while digging at an average depth of 23.44 feet.

Natoma No 3 dredge was commissioned in July, 1908, and is working on the north side of the river on what is known as Mississippi Bar. It is a close-connected-bucket elevator dredge, equipped with 61 8-cubic-foot buckets, dumping at the rate of 20 per minute.

Up to January 1, 1909, or during the first six months in operation, this dredge turned over 8½ acres of ground and handled 583,900 cubic yards of gravel, during 171 operating days, at a cost of 3.9 cents per cubic yard, while digging to an average depth of 42.5 feet in compact gravel.

Summary of Dredges Operating in the Folsom District up to January 1, 1909.

Company.	Dredge.	Construction of Dredge.	Date Began Operation.	Date Stopped Operation.	Time in Operation	Capacity of Buckets, cu. ft.	Cubic Yards Dredged.	Acres Dredged
Colorado Pac. Gold Drdg. Co.	Pacific No. 1.	Risdon† O. C.	April 1899	1903	yr. mo. 4	3½	960,000*	25*
	Pacific No. 2	Risdon O. C.	1902	1906	4	5	2,100,000*	54*
Ashburton Mining Co.	Ashburton No 1.	Bucyrus, D. L. C. C.	Mar. 1,1900	1906	7	7½	6,000,000*	155*
	Ashburton No. 2	Company C. C.	May, 1908		7	7	450,000*	8*
Syndicate Dredging Co.	Syndicate	Risdon O. C.	1901	1906	5½	5	3,000,000*	103*
Folsom Development Co.	Folsom No. 1	Bucyrus, W.E., C.C.	Feb. 20,'04	Aug.20,'08	4	6½	5½	2,995,000	54
	Folsom No. 2	Bucyrus, W.E., C.C.	Mar. 16,'04	Jan. 14,'09	3	9½	5½	5,350,000	150
	Folsom No. 3	Company, C. C.	Jan. 1,'05		4	8½	3,600,000	81½
	Folsom No. 4	Company, Bucyrus	Nov. 15,'05		3	1½	13	6,500,000	209
	Folsom No. 5	Bucyrus, W.E., C C.	Dec. 10,'05		3	½	9	3,350,000	39
	Folsom No. 6	Bucyrus, W.E., C.C.	Mar. 8,'08			9½	9	1,200,000	16.6
El Dorado Gold Dredging Co.	El Dorado	Risdon, O. C.	April 25,'05		3	8	7	1,260,000*	40*
Natoma Development Co.	Natoma No.1	Yuba C. C. C. C.	May 10,'08			7½	13½	1,850,000	60 02
	Natoma No 2	Yuba C. C. C. C.	Aug. 22,'08			4½	8	626,300	16.38
	Natoma No. 3	Yuba C. C. C. C.	July 2,'08			6	8	583,900	8.5

†C. C., Close connected: O. C., Open connected.
* Estimated; accurate data not obtainable.

Summary of Dredge Operations from 1899 to 1909.

Number of dredges	15
Acres dredged	1018.56
Yardage handled	37,125,200
Gross yield	$3,920,231

The foregoing gives six companies which were organized to operate in the district and were reduced to four at the end of 1908, owing to consolidations. Fifteen dredges were put in commission and 6 dismantled, leaving 9 in active operation at the end of 1908.

No. 169. Natoma No. 3 Dredge, showing construction of housing. Yuba Construction Company, builders. California type.

THE AMERICAN RIVER DISTRICT IN 1909 AND DURING THE FIRST HALF OF 1910.

On January 1, 1909, the Natomas Consolidated of California took over the holdings and equipment of three companies operating in the district. These were: The Folsom Development Company, five dredges, one machine shop, and one rock-crushing plant; the Natoma Development Company, three dredges and repair shops, and the El Dorado Gold Dredging Company with one dredge. The Natomas Consolidated also absorbed the holdings of the Natoma Vineyard Company. This reduced the number of companies operating to two, the Ashburton Mining Company remaining the same as in 1908.

The following table shows the dredging area belonging to the respective companies and the number of dredges operating in the field at the beginning of 1909:

Company.	Dredging Lands, Acres.	Dredges.		
		Working.	Idle.	Total.
Natomas Consolidated of California	5,621.66	7	2	9
Ashburton Mining Co.	205	1		1
Totals	5,826.66	8	2	10

During the year 1909 the above seven dredges of the Natomas Consolidated of California turned over 321.48 acres and handled 13,975,185 cubic yards of gravel at a cost of 3.6 cents per cubic yard, while digging to an average depth of 27 feet in ground ranging from 19 to 70 feet deep.

During the same period the four dredges of the Feather River division of the Natomas Consolidated of California turned over 79.7 acres and handled 3,555,552 cubic yards of gravel, during 26,597 operating hours, at an average cost of 4.9 cents per cubic yard, in ground ranging from 25 to 54 feet deep.

Table Showing Dredges Operating in the American River District in 1910.

Name of Company.	Total Dredges.	Name of Dredges.		
		Working.	Idle.	Constructing.
Natomas Consolidated of California	Natoma No. 1	Natomas No. 1		
	Natoma No. 2	Natomas No. 2		
	Natoma No. 3	Natomas No. 3		
	Folsom No. 4	Natomas No. 4		
	Folsom No. 5	Natomas No. 5		
	Folsom No. 6	Natomas No. 6		
	Folsom No. 3	Natomas No. 7		
	Natomas No. 8			Natomas No. 8
	El Dorado		El Dorado	
Ashburton Mining Co.	Ashburton	Ashburton No. 2		
Totals	10	8	1	1

Natomas Consolidated of California.—The Natomas Consolidated of California is one of the two largest gold dredging companies in the State, and as a combined dredging and reclamation company is the largest of its kind in California. It is a California corporation, with a

No. 171.   Natomas Crusher Plant No. 2.   Southern Pacific tracks in foreground.

capitalization of $25,000,000 divided into 250,000 shares of stock, par value of $100 each, and an authorized bond issue of 25,000 bonds, par value of $1,000 each. The personnel of the company is composed of San Francisco men, as follows: President, Eugene de Sabla; first vice-president, W. P. Hammon; second vice-president, Louis Sloss; third vice-president, F. W. Griffin; secretary and treasurer, A. E. Boynton;

general manager, Newton Cleaveland. The offices of the company are in the Alaska Commercial building, San Francisco, with division managers' offices at Sacramento, Natoma, and Oroville.

The company controls about 6,529.66 acres of proven dredging land, and operates twelve dredges. Of this 908 acres and four dredges are located at Oroville, Butte County, and 5,621.66 acres and eight dredges in Sacramento County. The principal holdings of the company lie in Sacramento County, where, aside from dredging interests, the company

No. 172. Interior of machine shop at Dredge, now Natoma.

operates two rock-crushing plants and controls some 8,333 acres of horticultural land in the American River district, a part of which belongs, however, to a subsidiary company, called the Natoma Vineyard Company. Elsewhere in the county the company has acquired some 70,000 acres of land which it is reclaiming and making habitable for settlement. The capital invested in dredging property amounts to over $5,000,000. During 1909 the dredging interests were divided into three divisions: the Feather River division, comprising four dredges at Oroville; the Natoma division, comprising the holdings taken over from the Natoma Development Company; and the Folsom division, comprising the holdings taken over from the Folsom Development Company. In 1910 the dredging interests were put under two divisions: the Natomas and Feather River divisions, comprising, respectively, the properties along the American River and those near Oroville. The names of all the dredges in the Folsom district were changed to Natomas (see preceding table), and two of the Feather River dredges were closed down, so that during the year 1910 the company has been operating two dredges in the Oroville district (see page 144) and seven in the Folsom district, a total of 9 dredges.

During the year ending December 31, 1909, the dredges of all divisions, which were the 3 Natoma, 4 Folsom, and 4 Feather River dredges, including a few months' experimental work with two obsolete machines, turned over 412.77 acres and handled 17,582,736 cubic yards at a cost

No. 173. Blacksmith shop at dredge near Folsom. Repairing large buckets and tumblers.

of 3.9 cents a cubic yard, the total operating time being 73,767 hours, and the average depth 26.4 feet in ground ranging from 18 to 70 feet deep. The bucket capacity of the dredges ranged from 5 to 13½ cubic feet.

Since the advent of the Natomas Consolidated of California in the American River district, dredging operations have received a new impetus in that field; the yardage handled in 1909, under the new management exceeds that of 1908. During the 12 months ending December 31, 1909, the three dredges of the Natoma division turned over 176.14 acres and handled 6,615,071 cubic yards of gravel at a cost of 3.02 cents per cubic yard, while operating during a total of 20,704 hours, or an average of 19.1 hours per day, and digging to an average depth of 23.3 feet in ground ranging from 19 to 51 feet deep. During the same year the 4 dredges of the Folsom division turned over 155.34 acres and handled 7,360,114 cubic yards of gravel at a cost of 4.2 cents per cubic yard, operating a total of 26,055 hours, or an average of 18.1 hours per day, and digging to an average depth of 29.4 feet in ground ranging from 18 to 70 feet deep. This record for the first year must be considered exceptional in view of the fact that during the first half of the year a great deal of time was lost in putting the various boats in working trim, as well as teaching the men in the various departments to keep a more exhaustive form of records than previously used.

Many improvements were made by the company during the year. The shops and offices at Dredge, a station now called Natoma, were enlarged and the offices of the Natoma division moved to this place.

During the year 1909 the company constructed and put in operation one large rock-crushing plant, at a cost of about $225,000, and towards the end of the year construction of a large dredge was begun. This dredge will be known as Natoma No. 8. It is designed by the engineers of the Yuba Construction Company, and is being constructed by the construction department of the Natomas Consolidated of California. The dredge will be of the close-connected-bucket elevator type, equipped with 13½-cubic-foot buckets, and will have a digging depth of 55 feet below the water-line.

The machine shop at Natoma is conveniently located, well equipped and large enough to economically handle and repair all but the heaviest parts of a dredge. A broad gauge spur track from the main railroad line runs through the yards making it possible to unload machine parts and supplies directly into either the shops or warehouses.

During the year 1909 the company employed the following average number of men per month in its dredging operations in the Folsom district:

Folsom Division Dredges.

General, including office force, dredge superintendent, division
  manager _____ 20 men
Floating crew _____ 20 men
Chinese and Japs _____ 16 men
Crew on four dredges _____ 44 men
Machine shop _____ 30 men

  Total _____ 130 men
Total pay roll per month_____ $12,500

Natoma Division Dredges.

General, including office force, dredge superintendent, division
  manager _____ 6 men
Crew on three dredges_____ 30 men
Floating crew _____ 15 men

  Total _____ 51 men
Average total pay roll per month_____ 6,000

  Grand total _____ 181 men $18,500

The working time of a dredge is twenty-four hours, divided into three eight-hour shifts. Each dredge is operated by nine men, one dredge-master, and one shoreman, making a total of eleven men with now and then additional men on day shift as special work may require. The prevailing wages are: For winchmen, 45 cents per hour, or $3.60 per shift; for oilers, 35 cents per hour, or $2.80 per shift; for shoremen, 25 cents to 34 cents per hour, or $2 to $2.72 per shift.

Aside from the regular crew there is the clean-up man and his assistants. The company employs one clean-up man for each division, who has charge of the clean-up crew, and under his direction the gold on the tables of each dredge is removed at regular intervals, usually once a week. He also has charge of the retorting of the amalgam and the melting of the gold, and because of this last duty is often known as the company's gold man.

Aside from actual dredging operations the company employs a large number of men in connection with its other interests in the district, which are divided into departments as follows: Orchard and vineyard department, water department, rock-crushing department, and shops and construction department. The employees in these various departments number 145, divided as follows:

Orchard and Vineyard Department.

Foreman and division superintendent	5 men
White labor	20 men
Chinese and Japs	38 men
Total	63 men

Water Department.

Total	11 men
Monthly pay roll, orchard and vineyard department	$3,182
Monthly pay roll, water department	636

Rock-Crushing Department.

Rock-crushing plant No. 1	33 men
Rock-crushing plant No. 2	38 men
Total pay roll per month	$5,500

A summary of the monthly average pay roll and number of men employed by the company in the various departments is as follows:

Department	Average Number Men Employed.	Average Monthly Pay Roll.
Dredging	181	$18,500
Rock-crushing plants	71	5,500
Water	11	636
Orchard and vineyard	63	3,182
Totals	326	$27,818

The company maintains cottages and boarding houses for their employees. They are located in the following places: Folsom division, 25 four-room cottages, size about 24 by 32 feet, and 8 five-room cottages; Natoma division, at Natoma, 5 four-room cottages 24 by 30 feet, and one boarding house 96 by 32 feet; at Sacramento Bar, 1 boarding house 34 by 120 feet; and at Mississippi Bar, 3 cottages 22 by 30 feet, and 1 boarding house 30 by 90 feet, making a total of 41 cottages and 3 boarding houses. The cottages are rented to employees at an average rental of $15 per month.

5.   CALAVERAS COUNTY.

Calaveras County, comprising an area of 1,000 square miles, or 63,360 acres, with a population of about 12,500 inhabitants, is the central of the Mother Lode counties, extending from the foothills of the Sierra Nevada Mountains into the great San Joaquin Valley.   San Andreas is the county seat, but the principal town is Angels Camp, with a population of 3,000.

The Mokelumne River separates the county on the north from Amador and the Stanislaus River on the south from Tuolumne.   On the west it is bounded by San Joaquin and Stanislaus counties and on the northeast by Alpine.   The Southern Pacific railroad has three branch lines

No. 174.   Isabel Dredge, September 14, 1909, Jenny Lind, Calaveras County, Cal.   Note steel bow gantry.

running to Valley Springs, Angels Camp, and Milton, respectively.   In 1908 the county had 412 miles of public roads, 800 miles of irrigating ditches, one electric power plant, and about 120 miles of electric power lines.

The principal industries of the county are mining, lumbering, and stock raising.   The famous Calaveras Grove of big trees is a great feature of the forest region of the country.   Mining in Calaveras is not confined to gold mining alone, for it was in this county that copper mining first began in California, and before the great deposits of Shasta County were developed Calaveras was the leading copper producing county of the State.

Calaveras County has been famous for its surface placers, especially during the days of the "long-tom" and "rocker."   The gold now produced comes mostly from deep quartz and gravel mines.   Gold dredging has only been carried on since 1904, and so far only in two localities. At the present time, 1910, there are three dredging companies in active operation—The Mokelumne Mining Company, Calaveras Gold Dredging

Company, and Isabel Dredging Company. The Butte Dredging Company of Oroville is this year removing the Butte dredge to property adjoining that of the Isabel Dredging Company on the south side of the Calaveras River. For Butte Dredging Company see pages 119 and 120.

Table Showing Mineral Production of Calaveras County from 1900 to 1908.

Substances.	1900.	1901	1902	1903.	1904
Chrome	------	------	------	------	$375
Clay	------	------	------	------	100
Copper	$150,585	$268,000	$251,062	$297,263	414,399
Gems	------	------	------	------	------
Gold	1,649,126	2,024,685	2,072,939	1,904,125	1,789,184
Lime	------	------	------	------	5,500
Limestone	------	------	------	------	------
Mineral paint	3,800	500	778	1,000	385
Platinum	------	------	------	------	------
Pyrites	3,583	------	------	------	------
Quartz crystals	18,000	17,500	------	------	------
Silver	80,762	44,687	46,234	68,280	65,611
Unapportioned	------	------	------	------	------
Totals	$1,905,856	$2,355,372	$2,371,013	$2,270,668	$2,275,554

Substances.	1905.	1906.	1907.	1908.	Grand Total.
Chrome	$300	$280	$840	------	------
Clay	300	50	250	$250	------
Copper	572,022	956,315	609,203	555,704	------
Gems	10,000	------	------	------	------
Gold	1,736,816	1,644,234	1,097,974	1,378,511	------
Lime	------	------	------	------	------
Limestone	15,430	7,635	16,955	31,446	------
Mineral paint	1,900	------	------	250	------
Platinum	------	250	------	------	------
Pyrites	------	------	------	------	------
Quartz crystals	------	------	10,000	10,000	------
Silver	------	74,099	54,420	62,727	------
Unapportioned	------	------	------	------	$50,075
Totals	$2,415,627	$2,682,863	$1,789,642	$2,038,888	$20,155,558

The Mokelumne Mining Company was incorporated under the laws of Delaware in 1901. The president and general manager of the company is David Pepper, Jr., Philadelphia, Pa., and the resident manager is William C. Calley, Wallace, Calaveras County, Cal.

The holdings of the company comprise an area of about 250 acres of dredging land located along the Mokelumne River in Calaveras, San Joaquin, and Amador counties at the junction of these three counties, about three miles from the town of Wallace.

The original operations of the company consisted of digging the gravel with a 90-ton Vulcan steam shovel, which dumped into a hopper. The gravel was then elevated by Robins belt conveyors to a stationary tower and dumped into a stationary sluice. This mode of operation did not prove successful, and in 1904 a bucket elevator dredge was built by the company on the ground.

The machinery for this dredge was made by Geo. V. Cresson & Co. of Philadelphia, and the equipment consisted in a general way of direct current motors driven by a coal and oil burning steam plant on shore. This plant and dredge was operated at irregular intervals until July, 1908, when the dredge was reconstructed and power brought in from the American River Electric Power Company's transmission line.

The dredge at present operating is a 5-cubic-foot close-connected elevator dredge carrying 74 buckets, and equipped with an independent Risdon ladder hoist and a Risdon 6-drum headline winch. The tailing stacker is a Robins belt conveyor carrying a 30-inch belt. The gold saving tables are of the double deck type and cover an area of one thousand square feet, the side sluices having an area of about 300 square feet. The pumps for supplying water for washing the gravel consist of one 7-inch and one 8-inch Krog centrifugal and one 3-inch high pressure pump.

The electric motor equipment installed upon the dredge consists of General Electric Company 60-cycle, 3-phase, 440-volt motors, distributed as follows: Main digging motor 100-horsepower V.S.; ladder hoist, 25-horsepower V.S.; main hoist, 25-horsepower V.S.; pumps 75-horsepower C.S.; high pressure pump, 5-horsepower; shaking screen and stacker motors, each, 20-horsepower C.S.

The normal digging capacity of the dredge is from 140 to 175 cubic yards per hour. The company contemplates building a new dredge in 1911. The company employs about twelve men.

The Calaveras Gold Dredging Company began operations in 1904. The president and manager of this company is S. A. Moss, 807 Alaska Commercial Building, San Francisco, Cal. The holdings of this company comprise an area of about 350 acres located in township 3 north, range 10 east, along the Calaveras River near Jenny Lind. The ground averages in depth about 33 feet. The gravel is coarse and in most places overlain by hydraulic tailing.

The dredge was constructed in December, 1903, by the Western Engineering and Construction Company and equipped with Bucyrus machinery. The hull of the dredge is 105 feet long, 36 feet wide, 7 feet deep and draws 4 feet six inches. The digging ladder is plate girder construction carrying 5-cubic-foot close-connected buckets. The tailing stacker is a Robins belt conveyor 90 feet long, carrying a 32-inch belt 187 feet long.

The electric motor equipment as installed upon the dredge has a rated capacity of 208-horsepower distributed as follows:

Pressure pumps, two 7-inch centrifugal	50 h.p.	850 r.p.m.	C.S.
Primary pump, one 2-inch centrifugal	3 h.p.	1700 r.p.m.	C.S.
Sand pump, one 6-inch centrifugal	30 h.p.	850 r.p.m.	C.S.
Shaking screen and stacker motor	30 h.p.	850 r.p.m.	C.S.
Bucket drive motor	75 h.p.	600 r.p.m.	V.S.
Winch motor	20 h.p.	1200 r.p.m.	V.S.

All motors are Westinghouse, 3-phase, 60-cycle, 400-volt. The company employs about twelve men.

The Isabel Dredging Company began operations in 1908. Until recently the personnel of the company consisted of Colorado mining men, with W. P. Bownbrights of Colorado Springs, president; H. J. Reiling, vice-president; A. Pollock, secretary and treasurer, San Francisco, Cal.; Fred J. Estep, resident manager, Jenny Lind.

It is understood that the company has lately been sold to California mining men, but nothing definite can be given as repeated requests for information have been refused.

The holdings of the company comprise an area of about 100 acres, located along the Calaveras River at Jenny Lind and adjoining the holdings of the Calaveras Gold Dredging Company. The ground averages in depth about 35 feet. The gravel is coarse and is overlain by a heavy clay in places several feet in depth and by a sandy loam. The overburden of clay and loam exceeds in places 10 feet in depth, and is said to have caused a great deal of difficulty in dredging.

The dredge on this property was originally built for the National Dredging Company, a Wyoming corporation, and later taken over by the Isabel Gold Dredging Company of Colorado Springs, Colo. The dredge was put in commission April 2, 1908, the hull being constructed by the Western Engineering and Construction Company and Bucyrus machinery installed by the National Dredging Company.

There are 215,000 feet of lumber in the hull, which is 100 feet long by 39 feet wide by 7 feet 8 inches deep with a draught of 5 feet. The dredge was built to dig 36 feet below the water line. The digging ladder is lattice girder construction and carries 5-cubic-foot close-connected buckets. The tailing stacker is a Robins belt conveyor 90 feet long between centers carrying a 30-inch belt.

When first built the dredge was equipped with a combination revolving and shaking screen; the revolving screen having perforations of large square holes about 3 inches by 5 inches. The gravel under 5 inches and sand would pass through the revolving screen and fall on to the shaking screens, which were placed directly below the revolving screen. The coarse material on the upper and lower screens passed out to the tailing conveyor and the very fine material, going through the lower screen, would pass over the gold-saving tables. The bow gantry of the dredge is entirely of steel construction.

In 1909 the dredge was reconstructed and the double screen arrangement discarded and many changes made, and it is thought that the difficulties encountered in handling and washing the clay and gravel will be effectively overcome.

The company employs about fourteen men.

6. STANISLAUS COUNTY.

Stanislaus County is situated at the head of the San Joaquin Valley, 100 miles southeast of San Francisco. From the ridge of the Coast Range, joining Santa Clara, it extends east to Tuolumne and Calaveras counties, and is bounded on the north by San Joaquin and on the south by Merced. The county has an area of 1,486 square miles or 965,900 acres. The population is given as 20,242, Modesto is the principal city, with a population of 4,800.

The San Joaquin River runs north across the center of Stanislaus

No. 175. La Grange Dredge, Stanislaus County.

County, and has the Tuolumne and Stanislaus rivers, which flow west from the Sierras, as tributaries. Water for irrigation on the east side of the San Joaquin River is supplied from the Tuolumne and Stanislaus rivers, the water being diverted by large dams. The Southern Pacific and Santa Fe main lines pass through Stanislaus, and two branch lines of the Southern Pacific skirt the foothills. An electric railway to Stockton, and one across the county from Oakdale to Newman, is projected.

Quartz mining has been carried on to some extent in the county, at La Grange; in the Sierra foothills, both quartz and dredge mining is being carried on.

The following table shows the mineral production of Stanislaus County, from 1900 to 1908:

Substances.	1900.	1901	1902	1903.	1904.
Brick					
Copper		$12,494	$18,676	$15,080	$931
Gold*	$21,212	15,700		52,869	50,000
Mineral paint	193	375	350	2,400	1,600
Platinum					20
Rubble					
Silver*				256	265
Unapportioned					
Totals	$21,405	$29,169	$19,026	$70,605	$52,816

14—GD

Substances.	1905.	1906.	1907.	1908	Grand Total
Brick	------	------	------	$7,000	------
Copper	------	------	------	------	------
Gold*	$50,000	------	$3,364	------	------
Mineral paint	2,125	$1,720	1,720	2,000	------
Platinum	------	------	------	------	------
Rubble	------	------	------	74,000	------
Silver*	240	------	28	------	------
Unapportioned	------	------	------	------	$82,317
Totals	$52,365	$1,720	$5,112	$83,000	$417,535

* U. S. Geological Survey reports gold and silver for Merced and Stanislaus counties together.

The La Grange Gold Dredging Company is the only dredging company operating in Stanislaus County.

This company began operations in 1908 and has one dredge at work. The president of the company is Jesse W. Lilienthal, San Francisco, Cal.; E. A. Wiltsee, New York, is general manager, and Archie Scott is resident manager at La Grange, Cal.

The holdings of the company comprise an area of about 200 acres of dredging ground, located along the Stanislaus River about four miles below the town of La Grange. The ground is said to average in depth about 35 feet. The gravel is medium coarse, and is overlain by a clay soil ranging in depth from 8 to 10 feet. The bedrock is soft, like at Oroville, and is easily dug by the dredge.

The ground was prospected by means of drills, and is said to

No. 176.   La Grange Dredge, June 6, 1909.   Note the cast steel bow gantry cap.

carry more gold than the average gravel at Oroville. The surface contour of the ground is flat and free from timber growth.

The dredge was designed by Geo. L. Holmes, San Francisco, most of the machinery being furnished by Meese & Gottfried Company, San Francisco, and Allis-Chalmers Company. In construction the dredge is similar to other standard California dredges. It operates against spuds and is equipped with 7-cubic-foot close-connected buckets, shaking screens, Holmes gold-saving tables and belt tailing conveyor, etc.

An independent motor is used for driving the bucket line and another motor for operating the ladder hoist. In the construction of the bow gantry the dredge differs from the usual type of dredges.

The company employs about fourteen men.

7. MERCED COUNTY.

There is only one gold dredging company in Merced County, the Yosemite Mining and Dredging Company, which began operations late in 1907, and has an operating plant of one dredge on the Merced River, near the town of Snelling. The general manager of this company is James W. Neill, Los Angeles, California, and the resident manager at Snelling is James H. White.

The holdings of this company comprise an area of about 400 acres, located along the Merced River, and consisting of river bottom and low land, periodically submerged and therefore not cultivated. The gravel is a clean wash, carrying practically no clay, and no boulders larger than can be handled by a modern dredge. It averages in depth to bedrock about 20 feet, and in places is overlain by a sandy loam several feet in depth. The gravel is quite loose and the water-level close to the surface. Attempts to reach bedrock by sinking shafts were unsuccessful, and the property was prospected by means of a Keystone drill.

According to the general manager's report the actual average of 27 test holes sunk over an area of 178 acres was 19½ cents per cubic yard, and the average from 50 holes and pits combined was 16.4 cents per cubic yard. On a tract of 113 acres 20 shafts were sunk to water-level and 26.7 cubic yards of gravel washed, which yielded 14.4 cents per cubic yard. On this same tract 16 drill holes gave an average of 23.4 cents per cubic yard contained in the gravel below water-level. Pits along the river bank are said to have yielded as high as 40 cents per cubic yard. The general manager estimates the cubical contents of this land to be 24,200 cubic yards per acre, the average value of the gravel 16½ cents per cubic yard, and the cost of dredging at 6½ cents per cubic yard. The gold is said to be relatively coarse, clean, and easily saved.

The Yosemite dredge was constructed from the Indiana No. 1 dredge, which was wrecked in the floods of 1907, while operating at Oroville. The hull of the dredge was constructed by the Yosemite Mining and Dredging Company, who also installed the machinery. See pages 115 to 117 for Indiana dredge.

No. 177. Yosemite Dredge, Merced County, California.

The electric power for operating the dredge of the Yosemite Dredging and Mining Company is supplied by a power plant owned by the company and located near the Yosemite Valley Railroad at a point about a mile and a half from the dredge. The house containing the generating machinery is a 30-foot by 60-foot building, made of corrugated iron nailed to a wooden frame of simple and inexpensive, though strong,

No. 178. Gold-saving tables of Yosemite Dredge, Merced County, California.

design. The power is generated by a 120-kilowatt, 2,200-volt, 3-phase, 60-cycle generator, driven by a 180-horsepower two cylinder gas engine, made by the Western Gas Engine Company. The engine is operated by No. 2 distillate, which is stored in a tank sunk in the ground outside the power house and near a spur from the railroad, so that a car load of distillate can be discharged into it through a short portable pipe. The distillate is pumped from the storage tank to a small tank inside the house and is then pumped to the engine. The amount of power required, measured at the power house, when the dredge is running, is about 140 horsepower. About 9,000 gallons of distillate are used each month. The engine consumes about one tenth of a gallon of distillate per horsepower-hour. The power is generated at 2,200 volts and is conducted to the dredge by No. 4 copper wire, and is there reduced by transformers on board to 400 volts, all of the motors being made for that voltage.

The power plant has given excellent satisfaction, few delays have been occasioned by it, and the cost of power is probably not over 1¾ cents per kilowatt-hour.

8. SHASTA COUNTY.

Shasta County, comprising an area of 4,050 square miles, or 2,590,000 acres, with a population of 25,000 inhabitants, is located at the head of the Sacramento Valley, and is one of the largest counties in the northern part of the State. It is bounded on the north by Siskiyou County, on the east by Lassen, on the south by Tehama, and on the west by Trinity.

Redding, the county seat, with a population of about 5,500, is located near the Sacramento River, on the main line of the Southern Pacific railroad, a distance of about 260 miles from San Francisco. The Sierra Nevada and Coast Range mountains, rising to an altitude of more than 5,000 feet, cover a large part of the county on all sides except the south; on the east are several prominent mountains, the principal one of which is Lassen, with an elevation of 10,577 feet. A large area of the county is heavily timbered with yellow and sugar pine and fir, 377,126 acres being in the Forest Reserve.

The most important of the many beautiful rivers of Shasta County are the Sacramento, McCloud, Fall, and Pitt, which, with their tributaries, afford valuable power resources as the flow of water is constant and the fall rapid.

In 1909 Shasta County had 1,015 miles of public roads, 133 miles of steam railroads, three electric power plants, 251 miles of electric power line, 190 miles of irrigating ditches, and 35,000 acres of land under irrigation. Among the principal industries of the county, aside from mining, are lumbering, horticulture, and stock raising.

No. 179. View of Sacramento River below Redding, Shasta County.

From a mining standpoint Shasta is the most important copper producing county in the State. The following table shows the mineral production of Shasta from 1900 to 1908:

Substances.	1900.	1901.	1902.	1903.	1904.
Brick	$12,000	$12,000	$12,250	$17,500	$15,000
Chrome	1,400	1,950	4,275	2,250	1,470
Copper	4,166,735	4,881,048	2,496,731	2,171,497	3,439,974
Gold	733,467	927,975	878,706	771,242	1,031,429
Granite	------	2,000	------	------	------
Iron Ore	------	------	------	------	------
Lime	17,850	12,960	12,500	10,800	10,500
Limestone	1,150	------	3,600	5,400	------
Macadam	------	------	------	1,500	------
Mineral water	5,784	7,644	7,645	12,000	------
Pyrites	------	------	7,005	5,500	------
Silver	635,640	891,994	306,887	203,991	399,660
Unapportioned	------	------	------	------	------
Totals	$5,574,026	$6,737,571	$3,730,049	$3,201,680	$4,898,033

Substances.	1905.	1906.	1907.	1908	Grand Total.
Brick	$14,000	$22,000	$33,000	$12,000	------
Chrome	300	1,200	5,200	5,600	------
Copper	1,688,614	4,338,121	5,568,873	4,642,976	------
Gold	684,952	819,144	791,997	1,131,832	------
Granite	------	------	------	------	------
Iron Ore	------	------	400	------	------
Lime	8,000	8,040	31,900	9,100	------
Limestone	3,600	32,960	30,761	80,000	------
Macadam	------	------	------	25,000	------
Mineral water	12,000	------	5,500	20,000	------
Pyrites	------	89,895	197,364	539,553	------
Silver	167,548	434,483	370,211	517,596	------
Unapportioned	------	------	------	------	$11,120
Totals	$2,579,014	$5,745,843	$7,084,706	$6,983,657	$46,575,699

No. 180. Dredging ground on Clear Creek, near Horsetown, Shasta County.

There is only one dredging company in Shasta County, operating a bucket elevator dredge at the present time, although there are said to be several thousand acres of dredging land in this county.

No. 181. Old type dredge, showing tail sluices and tail scow, Shasta Company's dredge before reconstruction. Clear Creek, Shasta County.

The ground along Clear Creek, on which dredging is being done, ranges in depth from 20 feet to 40 feet and in character the gravel is a fairly fine loose river wash, carrying few large boulders and not much

No. 182. Old type double-lift dredge being reconstructed to a single-lift. Shasta Company's dredge during reconstruction. Clear Creek, Shasta County.

clay. The contour of the ground is fairly even and not much covered by timber growth. The bedrock is mostly soft and in character much the same as that at Oroville. The gold is generally fairly coarse and is worth about $19 per ounce. The gravel is said to yield from 6 cents to 25 cents per cubic yard.

The Shasta Dredging Company began operations in 1906 with one bucket elevator dredge. The personnel of this company consists mainly of Pennsylvania men. President, E. T. Kruse; resident manager, T. G. Janney, Redding, Cal.

The holdings of the company comprise an area of about 700 acres of dredging land, located along Clear Creek near Horsetown, about nine miles southwest of Redding.

The first dredge installed by the company was of the old double-lift type, equipped with long tail sluices and tail scow, and a large pump for elevating the gravel. After being partly destroyed by fire, this dredge was reconstructed in 1908 by the Western Engineering and Construction Company. The tail scow and sluice were removed and replaced by wood gold-saving tables, shaking screens, and a 30-inch Robins belt conveyor for stacking tailing. At the same time a 10-foot section was added to the digging ladder, increasing the digging depth of the dredge to 28 feet. The upper tumbler and middle gantry were moved forward and the main drive reset. The location of the pilot house was changed and the winch levers were moved from a position directly over the bucket line to a more favorable location on the forward starboard side of the dredge.

Shortly after the reconstructed dredge was put in commission, June 15, 1908, it was almost destroyed by fire on August 7, 1908, through what was believed to be a defect in the transformers. It was rebuilt by the Yuba Construction Company and put again in operation in July, 1909.

This dredge, which is now working, was built according to the latest practice used in dredge construction in California. It is equipped with 5-cubic-foot close-connected buckets, belt tailing stacker, and spuds. The gold-saving tables are of the single bank type, and extend outside the housing on both sides of the dredge.

Elsewhere in Shasta County prospecting for dredging land is being carried on and a large tract of land has been proved. Over 5,000 acres of land along Clear Creek and the Sacramento River have been bonded by one company, and is now being prospected.

Prospecting is carried on principally by shaft sinking; a new method of timber lagging is being used. Some of the property was prospected by drills a number of years ago, and it was reported that fair values were obtained. The cost of dredging at that time was higher than at present, and it is anticipated that with present methods a large portion

No. 183. Remodeled double-lift dredge with tail scow and tail sluices to single-lift close-connected bucket dredge with tailing stacker. Shasta dredge, 1908.

No. 184. Shasta dredge in course of construction, near Horsetown, Clear Creek, 1909.

No. 185. Hull of placer mining machine, Clear Creek, Shasta County.

of the property will prove profitable. Conditions for economical working are very favorable, and should the results of prospecting prove of
sufficient value to warrant the installation of the largest types of
dredges, one of the largest dredging fields in the State will be opened
along Clear Creek and the Sacramento River.

William Desilhorst has, for many years, successfully operated a steam
scoop along Clear Creek; while the capacity of the scoop is small, the
results obtained prove that the gravel of this section carries considerable
gold. The reported cost of dredging with the Desilhorst scoop is from
20 to 25 cents per cubic yard.

Along the Sacramento River, in the vicinity of Redding, a number
of suction dredges equipped with centrifugal pumps for lifting the
gravel have been in operation. These machines have so far not been
a success in extracting gold from the gravel in payable quantities.

Pneumatic caisson dredges have also been operated on the river, but
owing to the limited capacity of these machines, and to the depth of the
ground necessary to be removed before reaching bedrock, it is a question as to their profitable operation.

9.   SISKIYOU COUNTY.

Siskiyou County, comprising an area of 6,078 square miles, or
3,870,720 acres, with a population of about 20,000 inhabitants, is the
largest county in the northern half of the State, and the fifth largest
in California. It is bounded on the north by the State of Oregon for
a distance of 80 miles, and on the east, south, and west by Modoc, Shasta,
Trinity, Humboldt, and Del Norte counties, respectively. Yreka, the
county seat, has a population of about 2,000 people, and is located at an
altitude of 2,620 feet on a branch line of the Southern Pacific Railroad,
a distance of seven miles from Montague, a station on the main line
about 380 miles north of San Francisco.

No. 186. Outcrop of bedrock near dredging ground at Scott River, Siskiyou County, California. Town of Callahan in distance.

Owing to the Sierra Nevada and Coast Range mountains coming together in Siskiyou County over two thirds of the area is covered by mountains and forests, giving Siskiyou the most picturesque expanse of mountains, canyons, and forests, together with a multitude of streams

No. 187. Hauling dredge machinery in the mountains. Revolving screen for Scott River Dredging Company.

which run southward to the Sacramento and northerly and westerly to the Klamath River, the principal mountain ranges are the Klamath, Scott, and Salmon, ranging in altitude from 2,000 to 14,500 feet.

No. 188. Framing timbers for a placer dredge in the mountains. Siskiyou County, California.

The agricultural interests are chiefly confined to the valley lands, which comprise about 1,500 square miles, located in the western, central, and eastern parts of the county, in Scott, Shasta, and Little Shasta,

McCloud, and Butte valleys, respectively. The principal river is the Klamath, which, with its tributaries, drains nearly the entire county, and furnishes abundant water for irrigation, power plants, etc., and not being navigable, furnishes a natural dumping place for placer mines. The principal industries of Siskiyou County are lumbering, mining, stock raising, and farming. The county has 1,100 miles of public roads, 204.16 miles of railroad, 155 miles of electric power lines, and eight electric power plants, 300 miles of irrigating ditches, which at present supply water to 33,000 acres. The total assessed valuation of property in 1909 amounted to $18,412,339.

The mining section, which contributes a large share to the prosperity and progress of the county, is located in the west half. The following table shows the mineral production of Siskiyou County from 1900 to 1908:

	1900.	1901.	1902.	1903.	1904.
Copper			$23		
Gold	$951,397	$886,043	906,989	$613,576	$892,685
Lead					
Lime					
Limestone					
Mineral water	45,000	175,000	187,500	50,000	50,000
Platinum					21
Rubble					
Sandstone					
Silver	13,986	6,408	233	22	1,230
Unapportioned					
Totals	$1,010,383	$1,067,451	$1,094,745	$663,598	$943,936

	1905.	1906.	1907.	1908.
Copper			$39	
Gold	$803,035		398,017	$504,156
Lead			140	183
Lime			1,000	1,680
Limestone			300	
Mineral water			36,250	80,000
Platinum	93			
Rubble			39,000	
Sandstone	1,250	$1,500	12,897	1,485
Silver	2,499		3,037	6,125
Unapportioned				$1,202,732
Totals	$806,877	$1,500	$490,680	$593,629	$7,875,531

Along the Klamath River and its important tributaries are large gravel deposits. Hydraulic mining has been most profitable, and formerly the gold from the gravel mines of Siskiyou County exceeded in value the annual output of all other gravel mines in California, but the gold yield from that source is now exceeded by a number of other counties. River-bed mining, by the use of wing-dams, has been carried on more extensively than in any other county, and lately gold dredging has met with success. In 1910 there were two gold dredging companies operating elevator dredges in Siskiyou County, the Siskiyou Dredging Company and the Scott River Dredging Company.

Scott River Dredging Company.—The Scott River Dredging Company began operations in August, 1908, and has an operating plant of one dredge. The company is incorporated under the laws of the State of Maine, the officers being as follows: President, Thomas C. Bishop; vice-president, E. E. Brownell; secretary and treasurer, F. S. Mayhew;

No. 189. Scott River Dredge, Siskiyou County, California.

307 Clunie building, San Francisco; managers, Brayton and Mayhew; superintendent at dredge, R. C. Specht.

The holdings of this company comprise an area of about 200 acres, located in parts of sections 7, 17, 20, and 21, township 40 north, range 8 west, lying along Scott River at the town of Callahan. The property was prospected by means of drills averaging about one test hole per acre. The gravel averages in depth about 30 feet, and in character is a coarse wash, carrying little sand and no clay. The bedrock is irregular and consists of a decomposed schist, causing some difficulty in dredging.

Scott River Dredge.—The Scott River dredge was put in commission August 5, 1908, and was the first large close-connected-bucket elevator dredge to be built in Siskiyou County. During the first twelve months in operation it turned over 7.5 acres of ground and handled 354.961 cubic yards of gravel, while digging to an average depth of 30 feet.

This dredge was built to dig 30 feet below the water-line, and is equipped with a double plate-girder ladder, 85 feet 5 inches long, carrying 72 7½-cubic-foot buckets, each weighing 2,300 pounds, and driven by a 125-horsepower motor at a speed of 18 per minute. The hull is

110 feet long and 35 feet wide, and differs in construction from other dredge hulls in California in that it is 9 feet deep at the bow and 5 feet at the stern. The gold-saving tables are of the Holmes system, having a lineal area of about 110 feet and a riffle surface of 960 square feet. The revolving screen is of the Risdon type, and is 33 feet 6 inches long with a diameter of 6 feet, having a total screen area of about 360 square feet. The tailing stacker is Link-Belt Company make, 90 feet long between centers, carrying a 32-inch belt 200 feet long. The equipment

No. 190.   Scott River Dredge in course of construction.   Siskiyou County, California.

of this dredge consists principally of Link-Belt Company machinery, some parts being Risdon make. The hull was constructed by the Western Engineering and Construction Company, who also installed the machinery. The dredge has a rated capacity of about 100 cubic yards per hour.

The electric motor equipment as installed upon the dredge has a rated capacity of 345-horsepower, distributed as follows: 10-inch pressure pump motor for the supply of water to screen and gold-saving tables, 85-horsepower, C.S., 720 revolutions per minute; revolving screen motor, 30-horsepower, C.S., 900 revolutions per minute; main digging or bucket drive motor, 125-horsepower, V.S., 600 revolutions per minute; stacker motor, 30-horsepower, C.S., 900 revolutions per minute. Starboard winch motor, 30-horsepower, V.S., 900 revolutions per minute; ladder hoist motor, 45-horsepower, V.S., 514 revolutions per minute. All the motors are General Electric Company 3-phase, 60-cycle, 2,080-volt. The Scott River dredge was closed down during the summer of 1910.

Siskiyou Dredging Company.—The Siskiyou Dredging Company began operations February 16, 1910, with one large close-connected-bucket elevator dredge. This company was organized under the laws of the State of Maine, and is capitalized for $200,000. The directors and officers are: Geo. J. Carr, president; J. J. Hamlyn, vice-president; J. C. Osgood, secretary and treasurer, Oroville, California; H. G. Peake and A. Starr Keeler.

The holdings of this company comprise an area of about 255 acres, located at a point 5 miles north of Fort Jones, extending for a distance

No. 191. Hauling dredge machinery in the mountains. Section of digging ladder for Scott River Dredge.

of about 2½ miles along McAdams Creek in sections 12 and 1, township 44 north, range 9 west, and in sections 6 and 31, townships 44 and 45 north, range 8 west. The property was prospected by means of drills and shafts, and of the total area 165 acres are considered proved dredging ground. The gravel lies on a decomposed slate bedrock, and averages in depth about 40 feet, and in character is a medium coarse, clean wash, carrying no clay. The value is estimated at about 15½ cents per cubic yard.

Siskiyou Dredge.—The Siskiyou dredge is a 5½-cubic-foot close-connected-bucket elevator dredge, constructed by the Yuba Construction Company, and equipped with Bucyrus machinery, and in every respect is an up-to-date machine. The company employs about fifteen men.

VI.   RECLAIMING DREDGED LAND.

1.   ROCK-CRUSHING PLANTS FOR DREDGE TAILINGS.

The utilizing of gravel tailings from dredge operations for economic purposes, thereby removing the rock piles and making it possible to replant orchards and otherwise reclaim the land, is a comparatively new industry and was first successfully carried out by the Folsom Rock Company, a subsidiary to the Folsom Development Company.

An experimental plant was built upon the recommendation of R. G. Hanford, the vice-president and general manager of the company, following his investigations in the latter part of 1905 and in 1906. These investigations, continued in the East during 1906, demonstrated that the Farrell crusher, manufactured by Earl C. Bacon, Havemeyer building, New York, was the most suitable for the work, crushers of this make being seen successfully handling large boulders on Long Island.

It was found necessary to size the gravel before crushing, each grade of material being sent to different crushers, and the smallest size to corrugated rolls, and it was also determined that storage bins of large capacity, from which the rock could be automatically loaded into cars, were an absolute necessity for economic operation.

It was soon seen that the $50,000 or $60,000 first proposed for this experiment would not be enough to erect a complete plant of sufficient size to be a commercial success, but it was not thought advisable to go to greater expense until the successful crushing of the boulders had been fully demonstrated; the first plant, was, therefore, incomplete in design, many changes being made as experiments were carried on, and before success was fully assured the investment had increased to approximately $150,000.

Rock Crushing Plant No. 1 having a rated capacity of 1,000 tons of crushed rock per day, was erected and operated at Dredge, California, in 1907.   A detailed description of the plant is as follows: The rock in the tailing piles to be crushed is excavated by a 45-ton steam shovel and deposited into dump cars, which are hauled to the incline at the crushing plant by means of a 10 by 14 dinky engine.   An electric 30-horsepower hoist located at the head of the incline hauls the dump cars by means of a steel cable to an elevation of 30 feet above the ground, where the gravel is discharged on to an inclined grizzly feeding into a 42 by 26 Farrell crusher.   The sand and small gravel passing through the grizzly is collected in a bin under the grizzly, and carried to the waste pile by means of a 20-inch belt conveyor 200 feet long.

The crushed rock from the 42 by 26 crusher is fed into a 24-inch bucket conveyor and elevated to a height of 60 feet above the ground, and discharged into a sizing screen 48 inches diameter by 24 feet long placed above the timber storage bins 60 feet long by 20 feet deep by 30 feet wide and located over the railroad tracks to feed the rock directly to the cars. The rejections from the screens are fed to a 36-inch by 10-inch Farrell crusher which discharges the recrushed rock into the 24-inch bucket elevator for redistribution by the sorting screens.

The gratifying success met with in crushing these hard round boulders, together with the great demand for the product, demonstrated

No. 192. General view of Folsom Rock-Crushing Plant No. 1. Began operations August, 1907.

so fully the advisability of increasing the storage capacity and revising the method of carrying the gravel from the dump cars to the crushers that it was decided to increase the capacity of the original plant, and to make the following additions and changes.

The electric hoist, the incline dump car track, and the flat grizzly were discarded and replaced as follows: A main receiving hopper was installed on level ground, directly under the narrow gauge tracks supporting the dump cars. The mouth of the hopper is 20 feet square and covered with timber grizzly bars, lined with ½-inch thick steel, spaced 8 inches apart. The outlet from the bottom of the hopper is controlled by a swinging door, and discharges directly upon a 36-inch Robins belt conveyor, erected at an angle of 16 degrees, raising the gravel to a vertical height of 32 feet above the ground and discharging into a 60-inch diameter by 16-foot long revolving screen, perforated with 4½-inch holes, having a dust jacket 84 inches diameter by 12 feet long, perforated with 3-inch holes, the gravel being separated into three sizes. The boulders over 4½ inches are fed to the 42 by 26 crusher; the boulders 4½ and over 3 inches are fed to a 36-inch by 10-inch

Farrell crusher, discharging its product on to a belt-conveyor, which empties its load "for distribution in the storage bins" into the 24-inch bucket elevator, common to all crushers. The 3-inch gravel and under is conveyed by an 18-inch belt-conveyor to a 42-inch diameter by 10-foot long screen, and separated into two sizes of gravel and fine material. This gravel is fed into two sets of 30-inch by 20-inch corrugated rolls, which discharge their product upon an 18-inch belt-conveyor, common to both, and carries same to a 24-inch bucket elevator, discharging its load into a 48-inch by 24-foot sizing screen, located over the storage bins.

No. 193.   Folsom Rock Crushing Plant No. 1.

The storage capacity was increased by installing two 12-inch horizontal Robins conveyors by 120 feet long directly under the bottom of the existing overhead timber storage bins to receive the extra material for storage, which is conveyed to a 20-inch horizontal Robins belt conveyor 200 feet long, erected 35 feet above the ground and running parallel with the railroad loading tracks. This 20-inch conveyor is provided with an automatic tripper which deposits the sized rock into six separate piles. Below each storage pile is a 6-foot wide by 6-foot high by 60-foot long tunnel constructed of heavy timber. Each tunnel is provided with three loading gates to feed the rock to be shipped on to a 20-inch conveyor which rises out of the ground at an angle of 18 degrees to a vertical height of 15 feet and discharges the rock into an inclined chute feeding into the railroad cars on the third track.

There are three separate spur railroad tracks, two passing directly under the overhead storage bins and the third runs parallel with the storage piles. The cars on all three tracks can be loaded at the same time.

The above additions to the original plant reduced the operating cost and increased the output of the plant, thereby increasing the shipping capacity to 1,500 tons per day.

The total equipment of No. 1 rock-crushing plant consists of:

No. 1 main sorting screen, 60 inches diameter by 16 feet long.
No. 2 sorting screen, 42 inches diameter by 10 feet.
No. 3 and No. 4 sizing screens, 48 inches diameter by 24 feet long.
1 Farrell 42-inch by 26-inch jaw crusher.
2 Farrell 36-inch by 10-inch jaw crushers.
1 Farrell 36-inch by 6-inch jaw crusher.
2 30-inch by 20-inch corrugated rolls.
2 24-inch bucket elevators.
14 Robins Conveying Belt Company's conveyors having a total length of 1,500 feet.
9 440-volt motors, ranging from 5 to 200 horsepower, respectively.
4 2000-volt motors, 50 horsepower.
2 2000-volt motors, 75 horsepower.
2 2000-volt motors, 100 horsepower.
1 2000-volt motor, 150 horsepower.

The total motors, when operating at full load, require 485 kilowatts.

The operations of Plant No. 1 demonstrated that the crushed product from the dredge tailings made the best material for road metal, asphalt macadam, macadam roads, and all kinds of concrete work. It was also shown to bind well with crude asphaltic oil. The suitability of the dredge tailings for the above purposes is due to the great proportion of basaltic rock in the gravel.

The following is a report of a laboratory test of the crushed product from Plant No. 1 made by Smith-Emery & Co., inspecting and testing engineers, at San Francisco, California:

Concrete Rock Test.

Character of rock --Basalt
Specific gravity ------------------------------------ 2.94
Hardness --- 5.
Toughness ------- ------------------------------------Very tough
Clay and dirt --- ------------------------------------None
Edges ----------- ------------------------------------Sharp with well-defined faces
Flaws and seams -------------------------------------None
Absorption --- .01
Voids ---.50 per cent

Screen Test.

Weight per cubic yard ------------------------------2,592 pounds

Rattler Test.

Charge: Dry rock ----------------------------------:-- 100 pounds
Speed: 28 revolutions per minute-----------------5,040 revolutions
Weight of rock in rattler at end of 5,040 revolu-
    tions -- 90 pounds
Loss by abrasion-------------------------------------- 10 per cent

Dust and Fines from Rattler.

Retained on 4-mesh ----------------------------------- 2 per cent
Retained on 10-mesh ----------------------------------- 0.5 per cent
Dust --- 7.5 per cent

Rock-Crushing Plant No. 2.—It was only a few months after the advent of the Natomas Consolidated of California in the field that work on rock-crushing plant No. 2 was commenced. This plant, which is located near Fair Oaks, or about 15 miles from Sacramento City, was completed and put in operation in July, 1909. It was designed to suit the peculiar conditions attendant upon crushing boulders, and is the result of the experience gained in the operation of No. 1 plant. The chief points sought were storage for large quantities of crushed material and low cost of production, possible only with continuous running.

The design and details of this plan were worked out by the Western Engineering and Construction Company, which also erected the plant. The plant has a rated capacity of 1,500 tons of crushed rock per day of ten hours; the general arrangement is shown in the various cuts accompanying the following description.

The rock in the tailing piles to be crushed is excavated by a 40-ton

No. 194.   General view of Natomas Crusher Plant No. 2.

Bucyrus oil-burning steam shovel, mounted on 60-pound rail, standard gauge, portable track, running parallel with the tailing pile. The rock from the shovel is deposited in a train of Koppel cars, each of 4-cubic-yard capacity, mounted on 40-pound narrow gauge portable track, laid in a loop to enable loaded and empty cars to pass each other without necessitating switching when returning to shovel. The cars are handled by means of a 10- by 14-inch oil-burning dinky engine.

The receiving hopper is directly under the narrow gauge track carrying the 4-yard dump cars. The mouth of the hopper is 20 feet square, covered with 6- by 8-inch timber grizzly bars, spaced 8 inches apart, and shod with ½- by 6-inch steel wearing bars. The sides of the hopper are built of 4- by 12-inch plank, lined on the inside with ⅜-inch steel plate. The outlet from the bottom of the hopper to the main No. 1 conveyor is controlled by a swinging door built of steel and operated by a chain attachment wrapped around a small hand-wheel drum, placed in conveyor pit to regulate the flow of gravel to No. 1 conveyor belt.

The entire content of the receiving hopper is discharged upon a 36-inch Robins conveyor, 166-foot centers, speed 125 feet per minute. The timber frame work carrying the conveyor is erected at an angle of 16 degrees to enable the conveyor to discharge its load into the main screen "A," which separates the gravel into 2¼-inch and less, 3-, 4¼-,

No. 195. General plan of Natomas Crusher Plant No. 2 at Fair

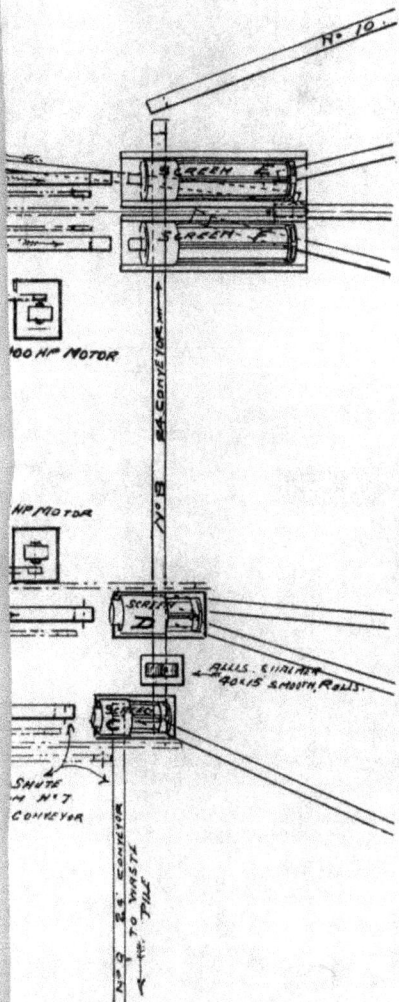

N° 10.

SCREEN E

SCREEN F

400 HP MOTOR

HP MOTOR

N° 8 CONVEYOR

B·X·B

SCREEN D

ALLIS-CHALMERS
40×15 SMOOTH ROLLS.

SHUTE
ON N° 7
CONVEYOR

N° 9 SEL. CONVEYOR
→ TO WASTE PILE

Oaks bridge.

No. 196. Natomas Crusher Plant No. 2, dinkey engine hauling dredge tailing to receiving hopper.

No. 197. Natomas Crusher Plant No. 2 in course of construction. Showing main conveyor.

No. 198. Natomas Crusher Plant No. 2, showing receiving hopper and main conveyor under cover.

and 8-inch sizes. The 3- and 4½-inch gravel is discharged into two 36- by 10-inch crushers and the 8-inch gravel into one 42- by 36-inch crusher. The ⅜-inch and sand are separated from the 2¼-inch gravel by screen "B" and discharged into two 40- by 20-inch corrugated rolls.

No. 199. Rolls for crushing dredge tailing used on Natomas Crusher Plant No. 2.

The product from the corrugated rolls is carried to screen "D" and separated into two sizes, ¾- and 1½-inch. The reject from screen "D" is fed into one 40- by 15-inch smooth roll and crushed to ⅜-inch and dust. These three products are carried to the storage piles by 20-inch conveyors. The product from the crushers is delivered into screens "E" and "F" and separated into ⅜-inch and dust; ¾-, 1½-.

No. 201. Conveyors for handling crushed rock. Natomas Crusher Plant No. 2.

and 2¼-inch are carried to the storage piles by 20-inch conveyors. The reject from screens "E" and "F" is recrushed by a 36- by 10-inch crusher and returned to screen "E" for redistribution.

In order not to have the whole plant closed down in the event of an accident to the rolls, conveyors Nos. 14, 15, 16, or screens "B," "C," and "D," a removable bottom, operated by rack and pinion, is installed in the receiving hopper at the head of No. 5 and No. 7 conveyors, together with an auxiliary chute, discharging to No. 9 conveyor. Thereby the 2¼-inch gravel from screen "A" is diverted from screen

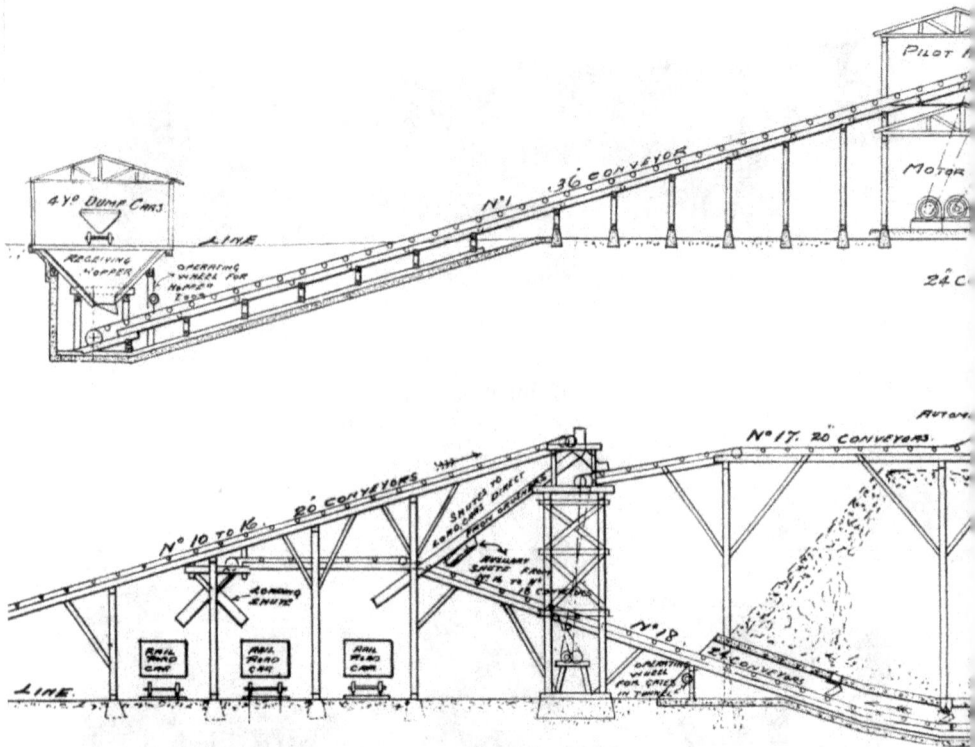

4 YD DUMP CARS

RECEIVING HOPPER

OPERATING WHEEL FOR HOPPER DOOR

LINE

N°1 36" CONVEYOR

PILOT

MOTOR

24 C

AUTOM

N°10 TO 16. 20" CONVEYORS

N°17 20" CONVEYORS

LOADING SHUTE

CHUTES TO LOAD CARS DIRECT

AUXILLARY SHUTE FROM 17 & 18 CONVEYORS

RAIL ROAD CAR

RAIL ROAD CAR

RAIL ROAD CAR

N°18 2" CONVEYORS

OPERATING WHEEL FOR CHUTE IN TOWER

LINE

No. 200. Section through Natomas

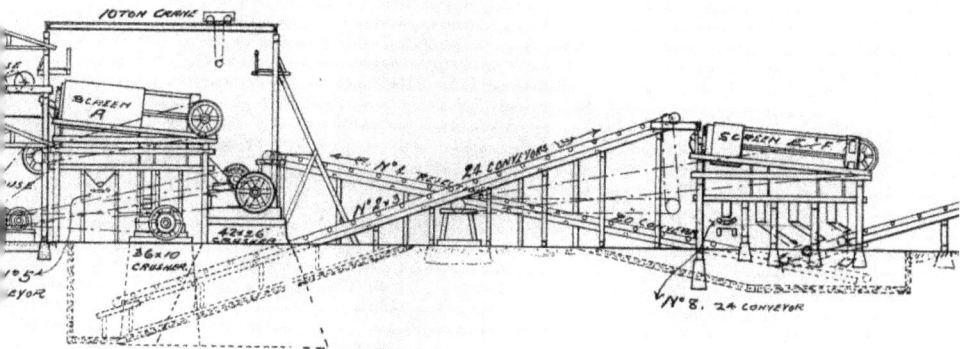

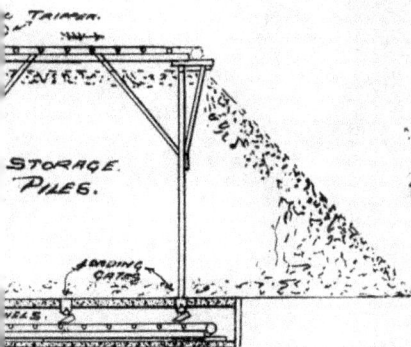

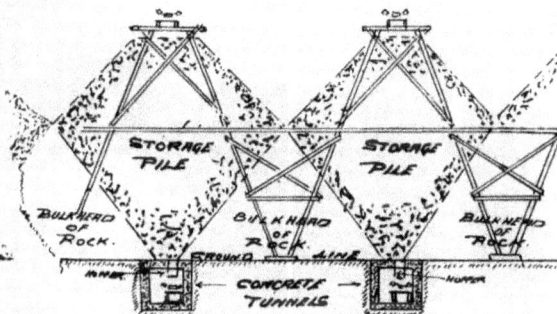

usher Plant No. 2 for dredge tailing.

"B" and delivered into the waste pile until necessary repairs have been made and the entire plant is ready for operations again. There are seven separate storage piles for crushed rock, each 100 feet long by 40 feet wide and 40 feet high. The size and class of material is as follows: 3/8-inch to dust, 3/4-, 1½-, and 2½-inch rock, all of which is an A1 product from the jaw crushers. The material from the rolls is a second-class product in sizes of 3/4- and 1½-inch rock. The seventh rock pile is composed of 3/8-inch round pebbles screened direct from the tailings pile without being crushed, and is used as roofing pebble.

Weight of crushed rock in pounds per cubic yards, according to sizes:

2½-inch size_____2,592 pounds per cubic yard
1½-inch size_____2,580 pounds per cubic yard
¾-inch size_____2,330 pounds per cubic yard
⅜-inch size_____2,400 pounds per cubic yard

Below the ground-line in the center of each of the seven storage piles is a reinforced concrete tunnel, 110 feet long, 6 feet wide, and 6 feet

No. 202. Ground storage piles of crushed rock at Natomas Crusher Plant No. 2.

high, inside dimensions, with roof and sides and bottom 12 inches thick. Each tunnel is provided with a 24-inch conveyor, 185-foot centers, operating at a speed of 250 feet per minute, which receives the material and elevates same on an angle of 18 degrees to a height of 25 feet, then running horizontally over the railroad tracks a distance of 25 feet discharges into a swinging chute, so placed as to direct the rock alternately into cars on separate tracks. Each tunnel is provided with four gates, on the swinging-door pattern, 20 feet apart, and so connected by levers and rods that they can be operated separately or together from a hand-wheel at the mouth of the tunnel.

The third track is loaded direct from the crushers by an inclined chute from the head of conveyors, Nos. 10 to 16, inclusive; at a suitable position in this chute is an auxiliary chute for the purpose of diverting the material, when necessary, on to No. 18 conveyors to load cars on two separate tracks from storage piles, and also to load cars on three separate tracks direct from crushers.

The total equipment of the rock-crushing plant consists of as follows:

1 screen "A," 72 inches diameter by 22 feet long.
2 screens "E" and "F," 60 inches diameter by 24 feet long.
3 screens "B," "C," and "D," 48 inches diameter by 10 to 12 feet long.
1 Bacon 42- by 26-inch jaw crusher.
3 Bacon 36- by 10-inch jaw crushers.
2 Allis-Chalmers 40- by 20-inch corrugated rolls.
1 Allis-Chalmers 40- by 15-inch smooth roll.
30 Robins Conveying Belt Company's conveyors having a total length of
      4,152 feet.
4 50-horsepower electric motors.
1 75-horsepower electric motor.
5 100-horsepower electric motors.
2 150-horsepower electric motors.

The motor equipment is housed in a main motor house 24 feet wide by 60 feet long, timber framing, and covered on the outside with No. 24 corrugated iron. The pilot house, 24 feet square, is mounted on the roof of the main motor house, giving the operator a clear view of the entire plant.

The switchboard and controllers are designed especially to meet the requirements of this plant. A separate and different colored light is installed directly over each controlling switch, and a duplicate light and auxiliary switch is placed close to each electric motor. Each motor has a push-button and an electric bell near the auxiliary controlling switch, and is designated by a certain number of rings on the bell. In case any part of the plant gets out of order the man in charge of that particular part of the plant signals to the operator in the pilot house, who immediately closes down all or any part of the

No. 203.   Natomas Rock-Crusher Plant No. 2, conveyor tunnel under storage pile in course of construction.

No. 204. Natomas Rock-Crusher Plant No. 2, loading crushed rock.

No. 205. Machinery for Natomas Rock-Crusher Plant No. 2.

No. 206. Switchboard for controlling all circuits at Natomas Crusher Plant No. 2.

machinery affected by the breakdown, and at the same time telephones
to the foreman of the repair gang, who should be in the repair house,
telling him exactly where to go to find the trouble. All electric wires
to the different motors from the pilot house are laid in underground
conduits, thus lessening the possibilities of fires from this source. Electric power is delivered to the plant at 22,000 volt, and is transformed
down to 2,300 volt, and distributed to all motors at 2,300 volt. The
transformers are in a concrete tank, installed below the ground-line and
completely covered.

A hand-operated, 10-ton, overhead crane, having a span of 40 feet,
and 30 feet clear headroom, is installed upon a timber structure 100
feet long, placed
over the crusher
machinery and ex-
tending up to the
repair house, where
supplies and extra
parts are stored.
The repair house is
24 feet wide by 48
feet long, equipped
with a complete set
of tools and ma-
chinery to carry
out necessary re-
pairs to the plant.

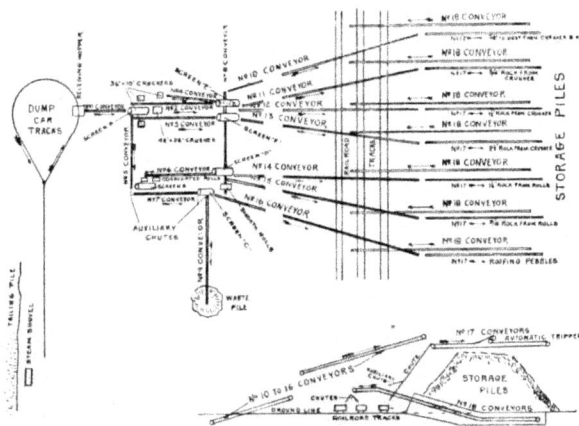

No. 207. Flow-sheet of Natomas Crusher Plant No. 2.

Two steel tanks for storing fuel oil, 11 feet 6 inches diameter by 10
feet deep, are mounted on a timber superstructure 25 feet above the
ground, and are placed midway between the railroad tracks and the
steam-shovel site. The oil pump is near the railroad track, and delivers
the oil from the cars into the storage tanks. The pipe line from the
storage tanks is laid underground, and is provided with ground-cocks
spaced every 80 feet up to the shovel site to supply fuel oil to the dinky
engine and steam shovel.

Two steel tanks for the storage of water, capacity 7,500 gallons, each
11 feet 6 inches diameter by 10 feet deep, are mounted on a timber
superstructure 37 feet above the ground and near the crusher site to
supply water to the shovel and railroad engines and also for fire purposes. The water is delivered to the tanks by one 6-inch deep well
pump, driven by a belt from a 50-horsepower motor. A most complete
pipe line for fire purposes has been installed, amply supplied with
hydrants and fire hose, placed in convenient places. A Fairbanks iron-
framed railroad track scale, with 50-foot platform, and a capacity of

No. 208. Line drawing of Rock-Crushing Plant No. 2 of the Natomas Consolidated of California. Began operation July, 1909.

200,000 pounds, is for the weighing of cars of crushed rock. The cars come down by gravity from the loading bins on to the scale platform.

Labor Required to Operate Plant.

1 superintendent.
1 steam shovel engineer.
3 steam shovel engineer helpers.
1 locomotive engineer.
1 gravel train brakeman.
4 laborers moving tracks of shovel.
2 laborers unloading cars at receiving hopper.
1 laborer regulating the receiving hopper door feeding the rock to the No. 1 conveyor.
6 laborers attending the four crushers and two rolls.
6 laborers attending to loading of cars at storage piles.
2 laborers placing sideboards in railroad cars.
4 laborers oiling conveyors and countershafts.
1 main operator in the pilot house.
2 electricians for motors.
2 mechanics in repair house.
1 clerk, weighing car.

The steam shovel takes eight minutes to load a train of seven cars of 4-cubic-yard capacity. The dinky engine hauls the train of cars to the

No. 209.   Valley Contracting Company's Rock-Crushing Plant at Oroville, Cal.   Erected by Newton Cleaveland and others.   Began operations August 12, 1908.

dump hopper, and returns the empties to the shovel in seven minutes. It takes ten minutes to load a 40-ton railroad car at the storage piles, and two cars can be filled at the same time with the same size of rock. There are seven separate piles of different sizes of rock; therefore, it is possible to load twelve cars per hour, making a shipping capacity of each size of rock of 480 tons per hour.

The economic value of dredge tailings depends largely upon their location. When near a railroad and within a reasonable distance from large cities or other markets there will be a market for this material for a period of many years. It is quite possible that, as the State of California becomes more thickly populated and railroad facilities increase, the demand for this material will rapidly exhaust the present available gravel piles built up by the gold dredges; as it is quite apparent that even at this early date, since the advent of the rock-crushing plants the dredge tailing has risen from a by-product to a co-product, commanding a place among the industries of the State.

No. 210. Steam shovel loading dredge tailing for Rock-Crushing Plant at Oroville, Cal.

2. REPLANTING DREDGED LAND.

The utilizing of dredged land for horticultural and agricultural purposes, etc., has been successfully tried in several places in California. In the Oroville district, James H. Leggett replanted many acres to eucalyptus, fig, orange, and almond trees and grapevines. The eucalyptus of both the "blue" and "red gum" variety and some fig trees are growing on unleveled tailing piles requiring no irrigation, fertilization, or special care. The three-year-old eucalyptus trees have attained a growth of from 16 to 20 feet; the four-year-old trees about 30 feet, and the tallest of the five-year-old trees a growth of about 40 feet. They range from 4 to 10 inches in diameter near the base. Nearly all of these trees were planted fifteen feet above the original surface of the ground, and when set out were about ten inches tall.

The grapevines, orange, almond, and fig trees, which were planted on leveled dredge tailing, are all growing rapidly and are in a healthy condition. Mr. Leggett says that the fruit from these trees ripens earlier and gives a better flavor than the fruit from trees growing in ordinary soil, and that owing to the heat retained by the rocks during the night, the danger from frost is a great deal less than on soil land.

No. 211.   Eucalyptus growing on unleveled dredge tailing in the Oroville District.

No. 212.   Eucalyptus and fig trees growing without irrigation on unleveled dredge tailing in the Oroville District, June, 1909.

He also states that it requires only five per cent of the water used in irrigation on ordinary soil to grow plants on dredged land.

The method of planting trees is as follows: A small shallow place is excavated in the rock tailing and filled with good soil, and the young trees, which are lifted with the usual ball of earth to the roots, are planted in this soil. In some instances old orange and other trees have been transplanted from land to be dredged to the tailing ground, in which case the method of planting is the same but more expensive. Usually the young plants require irrigation for some length of time,

No. 213. Orange trees and grapevines growing on leveled dredge tailing, Oroville District, July, 1909.

until the roots penetrate some distance into the moist zone of the gravel and sand.

It was found in dredging some land, adjacent to a tract of tailing ground, planted to grapevines, that the vines had grown roots in a few months, six to twelve feet long, extending down to permanent water level. Owing to the porosity and the moisture holding capacity of this ground alfalfa and grasses seem to do particularly well, as the roots of these plants quickly penetrate down to the permanent water-level.

One of the exhibits of products grown upon dredged ground, and displayed in the windows of the Chamber of Commerce at Oroville, showed a one-year-old peach tree, in a healthy, sturdy condition, having upon it a number of large well formed peaches. Some muscat grapes seemed to prove that a good production can be obtained from vines planted on dredged lands. An apple tree, dahlias, and hay completed this particular exhibit, which showed in a striking way the possibilities of reclaiming dredged land for profitable agricultural and horticultural purposes.

Successful results were obtained in the Folsom district with eucalyp-

No. 214.   Grapevines and almond trees on leveled dredge tailing, Oroville District, July, 1909.

tus trees. The Natomas Consolidated Company planted in May, 1909, several acres of leveled dredge tailing to eucalyptus trees, of the "blue gum" variety, which are doing well.

No. 215.   Eucalyptus trees planted on reclaimed dredging ground, Folsom District.
Three months after planting, June, 1909.

If tree planting continues to prove as successful as present experiments indicate, a large acreage of dredged ground will be useful for horticultural and agricultural purposes. The soil of most of the dredging lands in California was unproductive and was of little value for any purpose but mining. The only valuable land for other purposes was located in the Oroville and Folsom fields. In the Oroville field out of about 6,000 acres of dredging land, less than 400 acres were planted to fruit, and it is probable that not more than 1,000 acres of the remaining portions could have been used with irrigation, as it is ground of poor quality.

Much material has been made by antidébris exponents over the destruction by dredges of good orchards in the Oroville district and elsewhere, and it seems proper to state that previous to dredging most of the vineyards in the Oroville district were infested with the almost fatal phylloxera. Mr. Leggett among others was seriously contemplating taking up his vineyard, which comprises an area of 150 acres, and which is included in the above estimate of 400 acres of planted land in the district. The peach orchards also were not doing very well, most of them averaging but one crop in two and three

No. 216.   Orange trees growing on leveled dredge tailing in the Oroville District, July, 1909.

years, and many orchards were abandoned. Most of the ranches in the district were heavily mortgaged and it can not, therefore, be said that dredge mining has done much harm to this now prosperous locality.

In the Folsom district, as was also the case at Oroville, a large area of the dredging land had been worked by gravel miners previous to dredging, a large portion of the ground being left in such condition as to make it unfit for cultivation. Out of a total of about 6,000 acres of dredging land in this district not over 2,000 acres were planted to grapevines and for every acre turned over in dredging this land two acres of vines are planted on other land held for this purpose. Other conditions in this district were similar to those at Oroville previous to dredging.

On the Yuba River the question of reclamation is of another nature, most of the land being located on river bars covered by hydraulic tail-

No. 217. Grapevines growing on dredged land in the Oroville District.

ing and subject to overflow. The dredges in this district are erecting embankments to confine and control the flow of the Yuba River, this

No. 218. View of a large portion of the Oroville ground previous to dredging.

No. 219. View of ground in Oroville dredging field before dredging.

No. 220. General view of dredging ground in the Folsom District, previous to dredging, showing old placer workings.

No. 221. Dredging ground in the Oroville District.

No. 222. The gravel tailing ejected from lower sluices is covered by sand from upper sluices, facilitating the reclaiming of ground for farming purposes. This arrangement is not practicable in deep ground. Stanthrope, Queensland, Australia.

work being done on behalf of the State of California and the United States. Farm land being dredged in the vicinity of the Yuba River was very poor land, and will probably produce a better hay crop after the tailing is leveled than was possible before dredging.

Dredging ground located in Placer County on the Bear River contains some grazing land, but most of the land is subject to overflow and is covered by hydraulic tailing. In Calaveras County an area of 250 acres of dredged land has been sold to a farmer who is to receive the

No. 223. Apple and walnut trees, two years old, on Tewksburg Company's dredged land, near Bright, New Zealand. Soil, sand, and shingle are on the surface of the former land.

land in a level condition suitable for planting. The cobblestones were sold to a rock-crushing company and are being removed from the property. In Merced and Stanislaus counties the dredging ground was of little value for farming purposes, some of it being fair grazing land. In Shasta and Siskiyou counties the dredging lands were of no particular value excepting for mining purposes.

Attempts to utilize dredged land for agricultural purposes in New Zealand, seem, as a whole, to have met with success. In answer to inquiries by the California State Mineralogist the inspector of mines, E. R. Green, Wellington, New Zealand, writes as follows:

I have made inquiries from Messrs. McGeorge Bros., and am informed that flax growing was successfully tried on restored land, but owing to fluctuation on the market it was not likely to prove remunerative. Clover and timothy grass seed

No. 224. Dredged ground restored to city property. Oroville District, California.

sown on dredge tailing have been successful, and on the occasion of a visit to Waikaka by the Hon. Mr. McGowan we saw a paddock of 50 acres where the cattle were grazing on the dredge tailing thus treated, in preference to the original pasture.

Mr. John Turnbull, farmer, Waikaka, has informed me that he has sold land for mining up to $100 per acre, and that after dredging, by McGeorge's sluice-box method in restoring the soil as silt on the surface, he had rebought the land at the rate of $15 per acre and was glad to get it, and would continue giving that price, which, he was satisfied, results would warrant him in doing.

The ground so far worked by McGeorge's dredges at Waikaka has not exceeded 22 feet in depth, but now that gold has been found in the false bottom, McGeorge Bros. have a dredge almost completed to work at 40 feet from water level, so that it remains to be proved whether the soil can be successfully restored at that depth.

With regard to tree planting larch trees appear to have done the best. An area of four acres of larches planted on the Waikaka United claim seven years ago are now 10 feet to 20 feet in height.

As to the method of tree planting. The young trees are lifted with the usual ball of earth to the roots and planted out on the tailing in the ordinary way.

The following are abstracts from the papers and reports of the Minister of Mines, New Zealand. John Hayes, inspecting engineer (page 16, c. 3):

There has been an outcry in some quarters about valuable land being destroyed

for agricultural purposes by gold dredging operations.   This has undoubtedly been
the case in a few instances, but in others, where swampy marsh land has been
dredged, the effect has been to drain and sweeten it, and it is now growing sweet
grass and clover where rank sour grass and rushes grew before.   At the same time,
it can not be claimed that this land has been left in anything like so good a condition
as it might have been had advance stripping been practiced, and the soil and subsoil,
etc., deposited on the gravel tailing instead of them all being mixed up as at present.

At Waikaka, Southland, trees have been planted on the tailing left by one dredge
working on private land.   On my last visit to the locality I carefully inspected this
plantation, and found the young trees healthy and growing well.   The idea of
planting some of the tailing areas which were formerly swamp lands with native
flax suggested itself to my mind, and I submit this as offering a means of profitably
utilizing the ground from which the alluvial gold has been won.

No. 2 District, A. R. Campbell, inspector.

An Omeo dredge is making a good show in regard to resoiling, this being mainly
attributable to the depth and nature of soil dealt with.   This soil is stripped in
advance and sluiced down the box with the full body of water, but owing to the
tenacity of the material, a fair quantity of it stays on the tailing, the deposit being
in places 4 feet deep.   On this dredged ground there has this season been grown a
crop of oats on three acres besides small plots of vegetables.

That dredged ground can be successfully restored to town property
is evident as shown by the accompanying illustration No. 224 of the
house of O. C. Perry of Oroville, California.

This residence was built on ground that had been dredged and afterwards leveled, and although the house was built within a year after the
ground was dredged, there is no evidence of settling, and the plaster is
perfect in every respect, showing no cracks or flaws.

Mr. Perry considers dredged and leveled ground very desirable for
building sites, on account of the perfect drainage.   He has also found
that flowers and trees grow luxuriously with a shallow layer of top
soil over the cobbles.

VII. DEBRIS PROBLEM.

There has been a great deal of antidébris legislation in California, and many mining operations have been stopped by legislation to protect other interests. Recently there has been agitation against dredge mining, and the question raised as to what extent dredging pollutes and obstructs the streams. A committee appointed by the

No. 225. Feather River, near Gridley bridge.

board of an antidredge convention in reporting on the operations of the dredges along the American and Yuba rivers, found that on each of these rivers the dredges are doing no damage, and that instead of sending down any débris, much sand is being impounded and kept from going into the streams.

On the Feather River the same conditions exist regarding the present operations of the dredges, but in this district the early dredge operations left some obstructions in the river which are said to cause damage to adjacent low lands during flood times. As has been mentioned elsewhere, the present dredges in California, with the exception of those on the Yuba River, are all working inland, and hence the tailing stacks do not obstruct the flow of any of the rivers. The dredge cobbles in the Sacramento Valley have become valuable since the advent of the rock-crushing plants and it is not likely that any dredge tailing will remain very long where put by the dredges.

The question of pollution of streams by the spill water from dredge ponds is not a serious matter in California, as most of the dredges operate quite a distance inland from the rivers and practically all of the dredge water is impounded.

Before the water from dredge ponds was impounded, the question whether the colored spill water flowing into the rivers could have any injurious effect upon the river and farming interests was more frequently discussed in the Oroville than in other districts, and in an effort to obtain some exact information upon this point, the following samples were taken by H. Appel and W. S. Noyes at Oroville:

Sample No. 1. Outflow from under the tailing piles at the bank of the Feather River, near Pennsylvania Pond, about 10 inches in volume.

No. 226. Feather River below Oroville.

No. 2. About 100 yards down stream from No. 1, opposite a big riffle in the river, about 40 miner's inches flowing out from the cobble piles.

No. 3. Water crossing from east of Marysville road into the tailing piles west thereof, near the dam in the ditch on the Pennsylvania ground, and just south of the bridge crossing Gold Run.

No. 4. Outflow from pit of El Oro Dredge No. 1, about 15 inches in volume.

No. 5. Sample taken above the dam in El Oro No. 1 pond.

No. 6. Outflow from pond of El Oro No. 2 dredge.

All of these samples were sent to Abbot A. Hanks, San Francisco, with instruction to determine the foreign matter therein, in grains per gallon and in percentages by weight, and also to ascertain the fineness of the suspended matter. The following report on the foregoing was received:

SIR: I received from you recently six jars of water marked No. 1 to 6, inclusive. Each sample was mixed by careful shaking, and a measured quantity, one litre, drawn off. This was evaporated to dryness and the weight of the total solids determined as follows:

Mark.	Grains per U. S. Wine Gallon of 58415 Grains	Percentages by Weight.
1	187.783	.32 of 1%
2	215.772	.37 of 1%
3	274.092	.46 of 1%
4	753.468	1.29 of 1%
5	2029.466	3.47 of 1%
6	1014.733	1.73 of 1%

Experiments were then made to determine the fineness of the suspended matter in the water. In every sample all of this matter passed through a 200-mesh screen. This mesh is the finest product found in the market, and contains 200 apertures to the running inch. The finest cloth used in the manufacture of flour contains only 176 holes to the running inch, while the usual mesh is 156.

I believe that the only method of determining the exact size of the particles held in suspension in this water would be to measure them under a microscope with a standard form of micrometer. The suspended matter evidently consists of silt in its very finest form.

(Signed)   Abbot A. Hanks.

The solids shown in the above table included the solids in solution, which are always present in river water, and to determine these latter, two samples of the clear river water were later taken and their solids determined. They averaged 8.854 grains per gallon: equal to .0151 per cent. These amounts deducted from the corresponding items in the table leave the true amount of "suspended matter" in the water under consideration, as shown in the following table:

Mark.	Grains per U. S. Wine Gallon of 58415 Grains.	Percentages by Weight.	Solids in Solution Grains per Gallon.	Net Suspended Matter Grains per Gallon.	Per Cent Suspended Matter.
1	187.783	.32	8.854	178.929	.305 of 1%
2	215.772	.37	8.854	206.918	.355 of 1%
3	274.092	.46	8.854	265.238	.445 of 1%
4	753.468	1.29	8.854	744.614	1.275 of 1%
5	2029.466	3.47	8.854	2020.612	3.455 of 1%
6	1014.733	1.73	8.854	1005.879	1.715 of 1%

For the purpose of determining the size of the particles, a microscopical examination of the sediment was next made. Mr. Hanks' report on this follows:

Sir: In attempting to arrive at the size of these particles I have made a microscopical examination of the samples. Slides were prepared by evaporating small portions of each sample to dryness, and the sediment was measured both with a stage micrometer and an eyepiece micrometer. Under the microscope the sediment is seen to consist of an infinite number of particles of approximately the same size. These particles show a tendency to cake together on drying, but under a strong light it was possible to see the individual pieces and to measure them with a fair degree of accuracy.

I find the particles in both samples to vary in size from one one-hundredth of a millimeter to one four-hundredths of a millimeter. I have reduced these measures to decimals of an inch, as follows:

1/100 mm._____ .0003937 inches.

1/100 mm._____ .0000984 inches.

(Signed)   ABBOT A. HANKS.

This examination shows that the coarsest particles of the "suspended matter" in the discolored water flowing into the river (samples 1 and 2)

No. 227.   Mining on the headwaters of the Feather River, Phillbrook Creek, Butte County.
Bare boulders show the effect of flood waters.

were only about one tenth the size of flour particles and the finest but one fortieth of the size.

A careful consideration of the preceding will show that the material actually escaping into the river contained from .3 to .35 of 1 per cent of this extremely fine powder; or, in other words, but from 178 grains to 209 grains to the gallon.

In Australia some provision for the study and abatement of nuisances of this sort has been made in the creation of sludge abatement boards. The method of work by this board is not unlike that of the Interstate Commerce Commission; that is, complaints of nuisances as received are investigated by special agents, followed if need by public hearings, and finally, if the evidence justifies, by an order for its abatement.

It is an interesting fact that the same complaints regarding damages by dredges to farming interests and rivers are made in Victoria as have been agitated in California, and that there, as here, careful investigations prove the complaints essentially groundless.

In the reports of the Victoria board for 1908, in one case of alleged damage due to dredging, the board finds: "When a bucket elevator dredge was working a few years since in the bed of the Goulburn, some miles below Jamieson, settlers down stream were unaware of its presence. . We have found that local complaints are principally due to erosion of drains and creek banks."

The importance of sedimentation, due to erosion of ditches, roadways, and to other disturbances of natural conditions due to public works, is especially emphasized. In one case a complaining town was shown, by the testimony of its own engineer, to have furnished in this way to the stream more detritus than possible from the mining of which complaint was made.

The board further says: "Some of the farming witnesses object to all discoloration, and apparently would not be satisfied in summer time with water almost clear, though compelled even where there is no mining to use worse water during winter and spring. The board can not recommend the elimination of a valuable industry on account of ill grounded objections of persons not acquainted with the nature of the work, many of whom would not, owing to distance from the mines, be subject even to discoloration of water by their operation."

The whole work of the Victoria board seems to indicate that when problems like these are intelligently and honestly studied, the amount of damage is neither so large nor the means of abatement so difficult as to cause serious uneasiness.

VIII.   OTHER DREDGING FIELDS.

Outside of California, gold dredging is being carried on in the United States of America in North Carolina, Colorado, Montana, Idaho.   In the states of Oregon and Nevada there have, so far, been no extensive dredging operations.   One dredge is reported in southern Oregon, but is not thought to be a financial success.   Some years ago an attempt was made to dredge ground near Sumpter, in eastern Oregon, but with poor results.   The inferior construction of the dredges used may partly account of the nonsuccess of these operations.   In Alaska, considerable gold dredging is being done, and some dredges are operating success-fully in the Philippine Islands.

In foreign countries gold dredging is or has been carried on in the following places: Canada, Klondyke region, and British Columbia; Mexico, in the states of Sonora and Sinaloa.   In Central America, in the State of Honduras, considerable prospecting for dredging land is being done at the present time.   In South America in the following states: Colombia, French and Dutch Guiana, Brazil, Argentine Repub-lic, Chile, Peru, Bolivia, and Ecuador.   In Australasia, Victoria, New Zealand and New South Wales.   In India some dredges are operating in Burmah.   In Africa considerable prospecting for dredging land is being done on the west coast.   In Europe a number of dredges are operating in Siberia and the Urals, and some gold dredging operations have been reported from Servia.

A fuller description of dredging operations in some of these states and foreign countries is given in the following pages.

*GOLD DREDGING IN MONTANA.

The gold dredging industry owes a great deal to Montana.   Many of the early problems in gold dredging were solved by the operators at Bannock, where dredging was carried on as early as 1894.   At present, some of the largest and most modern dredges in the world are being operated at Ruby.   It is well worth while for those interested in gold dredging to visit Ruby, where, aside from seeing a number of features other than those seen in different dredging fields, a great deal of useful information can be obtained.   The dredge operators at Ruby, though isolated, and not in contact with other dredge operators, show little provincialism, and are generous in giving information to strangers.

The principal dredging field in Montana is located at an altitude of about 5,200 feet above sea level, in Ruby Valley, Madison County, at a

*A portion of this article is taken from a paper by J. P. Hutchins in the Engineer-ing and Mining Journal.

No. 228. General view of Ruby Valley dredging field. Note covered stacker on Conrey No. 1 Dredge in foreground.

place just below where Alder Gulch leaves a narrow canyon in breaking from the mountains. There are two dredging companies in the district, the Conrey Placer Mining Company and the Poor Farm Placer Mining Company. Both companies are under the same general management and have the same stockholders. Harvard University inherited from the late Colonel Gordon McKay a large interest in both companies, and it is estimated that during the life of the mines Harvard University will have profited to the extent of from $1,500,000 to $2,000,000.

Alder Gulch was one of the rich placer deposits discovered in the early sixties. In 13 miles of its course, over an average width of about 300 feet and an average depth of about 20 feet, it has produced a gross yield estimated by various authorities at from $75,000,000 to $150,000,000. The expensive methods of rocking and hand-sluicing material, which was mined by open-cutting with hand shoveling (often with ground sluicing of overburden), when the material was shallow, or by drifting and hoisting to the sluices when it was deep, were used. Attempts have been made to work parts of the gulch gravel with various types of mechanical excavators, but not with success.

The gravel of Ruby Valley ranges in depth from 30 feet to 60 feet, the depth increases with the slope of the bedrock, in a westerly direction. In character the gravel is coarse and compact, carrying, in places, considerable sticky clay and many large boulders. The bedrock is an extremely tenacious clay; varying in color from light cream to dark brown, and seeming to owe its origin to deposition of tufa and volcanic ash in slow moving or still water. It resembles the Oroville bottom, but is more tenacious and sticky, although generally smooth, with but gentle undulations. There is no particular concentration of gold on or near the bedrock; these features and the circumstance that when any considerable volume of it is excavated by the buckets it must be pried or dug out of them (for it will not discharge) make it not only unnecessary, but undesirable to attempt more than to skim the bedrock very slightly.

There is a marked contrast between the bottoms of Alder Gulch and its debouchment. Not only are they different in that one is said to be rough, "true" bedrock, with deep crevices, while the other is a smooth "false" bedrock without cracks, but in the gulch a marked concentration and penetration of gold have been noted, while absence of these features is observed in the dredging area.

The Ruby Valley gravels were prospected by means of drills and on part of the property the test holes were put down 150 feet apart and staggered on lines 200 feet apart, while a large portion of the property was divided into rectangles 330 by 660 feet and test holes sunk at the corners. The gold is 840 to 860 fine. Its coarseness varies; about 40 to 50 per cent passing through a 60-mesh, and about 15 to 30 per cent passing through a 100-mesh screen. Much of the gold is subangular and is coarse in comparison with the gold of the California dredging fields. It resembles the gold from Alder Gulch, showing the same relation to it as observed elsewhere between gold of debouchments and that of the creeks above them. The average gold content is about 25 cents per cubic yard.

Attempts to work the gulch gravel with mechanical excavators other than bucket elevator dredges did not prove successful. One of the early devices used consisted of a cableway to which was attached a

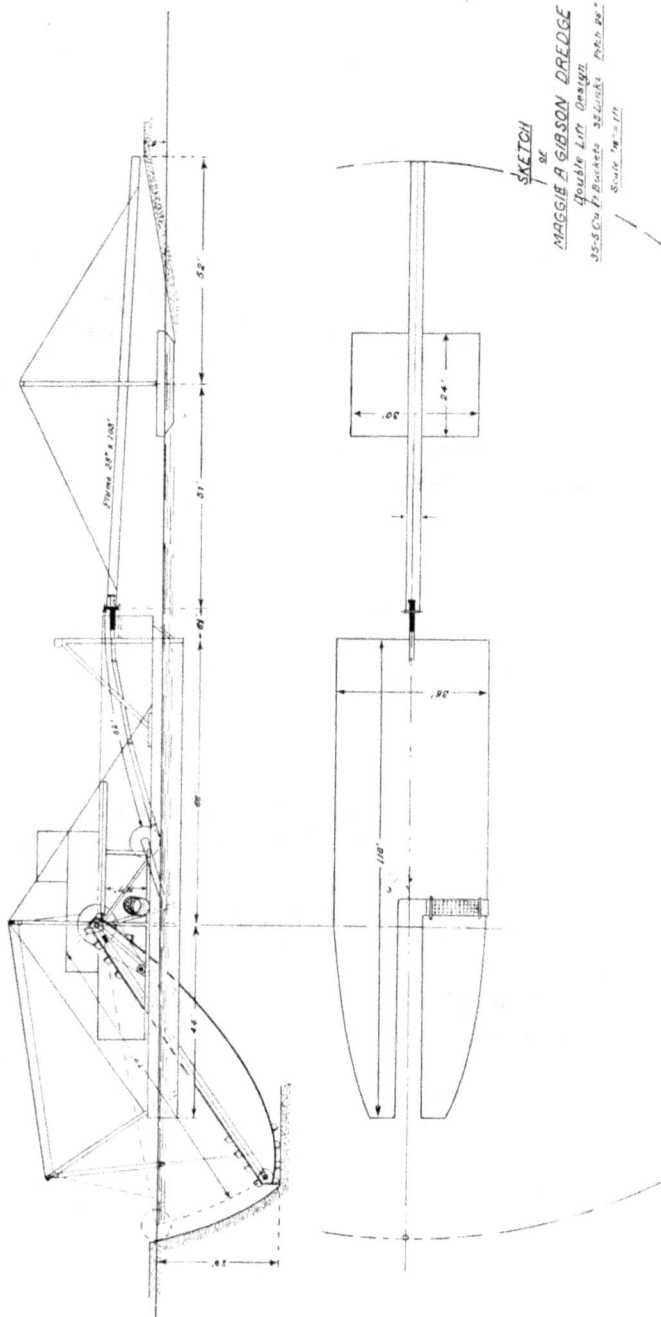

SKETCH
of
MAGGIE A GIBSON DREDGE
Double Lift Design
35-5 Cu Ft Buckets 32 Links Pitch 26"
Scale 1/8" = 1 ft

No. 229. Early Montana type. This dredge operated at Bannock, Montana, and was later removed to Ruby, Montana, where it operated five years on ground belonging to the Conrey Placer Mining Company.

bucket or drag scraper, that excavated the gravel as it was dragged across the cuts. The full bucket was trammed on the cableway to a stationary sluice mounted on a structural frame high enough above the ground to allow of grade and dump room for the tailing.

The plant, requiring over 30 men to run it, and costing about 30 cents per cubic yard to operate, was too expensive, and handling only about 500 cubic yards per 24 hours, was too limited in capacity. Like numerous other placer mining plants of but slight mobility, resultant low capacity and high cost prevented successful operation.

The first attempts to dredge the deposit included an excavator with an orange-peel bucket. The excavating machinery was mounted on a large car and track to permit movement; and a long boom was installed to enable discharge of the material from the bucket into a high hopper and washing machine, also mounted on a car. This installation was a failure; while possessing the many faults common to placer mining machines mounted on cars, it was also incapable of excavating the tenacious gravel. Attempts to work it were soon abandoned.

The first bucket elevator dredge in the district was the "Maggie Gibson." This machine was of the double-lift type, equipped with tail sluices and tail scow. It was first successfully operated at Bannock, Montana, and was reërected in Ruby Valley, and put in operation on leased ground belonging to the Conrey Placer Mining Company, where it operated for five years until the lease expired. The general equipment of the dredge consisted of 5-cubic-foot open-connected buckets, weighing each 600 pounds. The upper tumbler was driven by a sprocket wheel, and the tumbler shaft was 8 inches in diameter. The material after being raised by the buckets to a height of about 17 feet above water line, passed through a chain-driven revolving screen, having 4-inch by 5-inch openings. The fine material and gold going through the screen openings was collected in a sump in the hull of the dredge, from whence it was elevated by a 12-inch centrifugal pump to a sluice about 80 feet long, extending a distance behind the dredge. The coarse gravel and boulders passing out through the screen were dumped in the pond on the side of the dredge. Wooden spuds 48 inches by 48 inches were used to hold the dredge in place while digging.

In 1899 the Conrey Placer Mining Company installed a single-lift, 6-cubic-foot, open-connected-bucket elevator dredge, equipped with tail sluices and tail scow. The material was elevated to a height of about 30 feet above water line and dumped into a revolving screen, having 5-inch openings. The coarse gravel and boulders passing through the screen were dumped in the pond on both sides of the dredge by means of chutes, the fine material and gold passed to a sluice having a 7 per cent grade. This sluice was 140 feet long, 52 inches wide, and actuated at the stern, the outboard part 120 feet long being sustained on an auxiliary scow. The sluice was divided into two parts by a longitudinal partition (to allow continuous running in one half of it while the other half was being cleaned up). This partition was removed principally because unequal distribution of material on the two sides of the partition, due to listing, gave poor results.

The hull of the dredge was 100 by 44 by 7 feet, with a draft of 5.25 feet. It was equipped with three 70 horsepower boilers, one 14-inch and one 10-inch centrifugal pump lifting 24 feet. The 6-cubic-foot buckets, weighing each 700 pounds, were of weak construction and were later replaced by 7½-cubic-foot buckets weighing each 1,400 pounds.

A second dredge was equipped with 10-cubic-foot buckets. 200,000 feet of lumber, board measure, were used in the construction of the hull, which was made 100 feet long, 44 feet wide, and 7 feet deep, with a draught of 5 feet. The tail sluice was 4 feet 3 inches wide and 2 feet 4 inches deep. Two 100-horsepower boilers were mounted on the hull. The Conrey Company later discarded these two dredges and substituted for them two modern electric driven dredges, built by the Marion Steam Shovel Company.

In September, 1906, the Poor Farm Placer Mining Company put in operation a 15-cubic-foot open-connected-bucket elevator dredge. The hull of this dredge was made 130 feet long, 48 feet wide, 7½ feet deep, 400,000 feet of lumber being used in its construction. The digging motor has a rated capacity of 150 horsepower, and is located on a steel superstructure directly behind the upper tumbler. This motor is geared to the upper tumbler shaft through two intermediate shafts, a slipping friction and a circuit-breaker set at about 300 horsepower. This dredge was equipped with a tail sluice 66 inches wide and 135 feet long, being on a 7 per cent grade, and paved with strap longitudinal and transverse angle-iron riffles. Clean-ups on all the dredges are made every ten days. This dredge was designed by the late Julius Baier, former general manager for the company. Its average record for the past three years, up to 1910, has been 90,000 cubic yards per month.

In 1907 the Conrey No. 1 dredge, with 7½-cubic-foot buckets, was reported to dig an average of about 1,800 cubic yards per day; No. 2 dredge, with 10-cubic-foot buckets, 2,300 cubic yards per day; and the Poor Farm dredge, with 15-cubic-foot buckets, an average of 3,300 cubic yards per day at a cost of about 10 cents per cubic yard for the Conrey dredges and 6½ cents per cubic yard for the Poor Farm dredge, including depreciation, salaries, general expenses, etc. While these yardage figures seem small as judged by Californian standards, it must be remembered that there is a difference in the character of the respective gravels.

As the Conrey Company does its own repairing, etc., it has installed

a shop having, besides the common tools, a 250-ton hydraulic press.
In the gold dredges at Ruby, experience with the buckets, tumblers,
and ladders has been similar to that of other dredging districts, and
from the light buckets first used were evolved the later 7.5-cubic-foot
bucket, weighing 1,400 pounds, with 4⅝-inch pins, the 10-cubic-foot
bucket on No. 2 dredge, weighing 2,100 pounds, with 4¾-inch pins, and
15-cubic-foot bucket on the Poor Farm dredge, weighing 2,800 pounds,
with 5⅜-inch pins. Manganese steel pins and bushings were first used
and various other steel alloys have been tried; but now locomotive-tire
steel pins and manganese steel bushings are used with good results.
Complete spare bucket chains are kept on hand and each chain is considered and treated as a unit in repairing and renewing. This method
has for its object maintaining a uniform bucket pitch in all the elements
of the chain and thus preventing excessive wear on the bucket bottoms
and tumbler faces by slipping, as results when the uniformity of pitch
is destroyed. Bucket pins and bushings have been designed and used at
Ruby so that their life bears a certain relation to each other, thus permitting systematic methods in renewals.

Some experimenting was carried on with close-connected buckets on
the 7.5-cubic-foot dredge. The opinion is that the close-connected
buckets will dig with less power per cubic yard excavated, will dig more
per unit of weight of the digging apparatus, will dig with less surging,
will dig with less wear and tear per cubic yard excavated, and will thus
be more economical and reduce operating cost. Indicator cards also
showed an increase of about 17 per cent in the power required to dig
the indurated bottom gravel over that needed in loose top material.

The Poor Farm dredge has an 8-cubic-foot close-connected-bucket
line which is being used part of the time. New close-connected-bucket
lines of 9½-cubic-foot capacity are about to be installed on the Conrey
and Poor Farm dredges, one such bucket line being already in use.

Teeth of numerous types, with different shapes and angles of the
biting edges, have been tried, but all have been discarded after a great
deal of experimenting, as it was found that they gave no noticeable aid
in excavating. Bucket rollers of various materials have been tried,
manganese steel being one of the alloys; white iron is now used. Bucket
lips of cast manganese steel are used; those for the 15-cubic-foot
buckets weigh 500 pounds each. Considerable experimenting has been
done with bucket castings; in one instance annealed and unannealed
castings were used in the same bucket chain; no difference in wearing
quality and strength was distinguished.

A successful means has been evolved for keeping the lower tumbler
boxes lubricated. Protecting rings on the tumbler boxes, so arranged
as to take the side thrust of the tumbler while side feeding, have
angular spaces which are packed with flax packing; the outer ends of

the boxes are capped. Engine oil is used with good results. The bearings of the rollers on the bucket ladder are also protected and lubricated in a similar way.

A device for preventing the loss of material spilled from the buckets is in successful operation on all the dredges. The two sides of the ladder are connected on the bottom faces by a sheet and a stream of water from the upper end keeps this clean. Material washed down is discharged on both sides of the buckets about to round the lower tumbler by a device similar in shape to the roof of a house, and at such a point as to insure its being picked up in feeding laterally.

The save-alls, also called "deck sluices," save about 5 per cent of all the gold recovered. This again shows the result of the clay content preventing easy discharge. Wood was originally used; coal is had at the dredges for $4.90 per ton, and electricity at about 1 cent per kilowatt hour is available. About 15 tons of coal were burned per day per dredge.

It has been found that electric power is superior to steam by comparing the working of the new dredges with that of the old ones. It has a steadier, surer pull, with much less stalling action, and but slight variation in the chain speed in hard digging is noted. The buckets seem to fill more satisfactorily and there is much less racking, jerking and jarring. The ammeter of the digging motor fluctuates between 130 and 150 kilowatts. It may be said that, from the mechanical and economical view points, electricity is superior to steam in the dredging operations at Ruby.

The Conrey Mining Company has now two 7½-cubic-foot bucket elevator dredges, electrically driven, constructed by the Marion Steam Shovel Company. These dredges are of the California type, equipped with gold-saving tables and tailing stacker. It is said that as far as can be estimated, the percentage of gold saving is about the same as on the sluice boats, and the operating costs of the two types of dredges, character of ground being considered, is about the same. The dredges have averaged 82,000 cubic yards per month, and during the year 1909 the working costs and general management has been about 6¼ cents per cubic yard. The repair and replacement charge is about 40 per cent of the total expense.

A general description of one of the above dredges, which is shown in the illustration on page 254, is as follows: Digging depth below water line, 27 feet; length of bucket ladder, 72 feet; number of buckets, 60; capacity, each 7½ cubic feet; pitch of buckets, 32¾ inches; diameter of bucket pins, 4⅝ inches.

The hull is 98 feet long 9 feet deep and 44 feet wide, having a deck surface of 50 feet wide. The draught is about 6 feet. The center of upper tumbler shaft from deck is 25 feet 9 inches. The screen is of the telescope type and is 35 feet 5 inches long, 6 feet in diameter at the upper end and 4 feet 3 inches in diameter at the lower end.

Length of stacker 90 feet between centers. Width of stacker belt 32 inches. The stacker, as can be seen in the illustration, is covered over, in order to protect the belt, rollers and roller bearings from the cold weather in winter.

The Marion Steam Shovel Company is now constructing a 15-cubic-foot bucket elevator dredge for the Conrey Mining Company, which will be put in operation in 1910. This dredge is designed for a capacity of 300,000 cubic yards per month. There will be eighty 15-cubic-foot buckets in line, weighing each 4,200 pounds. The stacker will be 138 feet long and the spuds weigh 45 tons each. The motor equipment of the dredge will have a rated capacity of 1,050 horsepower.

It is expected that the cost of dredging with the new 9½-cubic-foot buckets will not exceed 5½ cents per cubic yard and with the new 15-cubic-foot bucket dredge about 4½ cents per cubic yard.

The Ruby dredges are the first to run throughout the winter in such a cold climate. Formerly it was thought that a dredge must be shut down about the time other placer mining operations ceased, and it was because of this belief, rather than any other reason, that dredges in many instances were idle during four or five months each year.

It is not an easy matter to operate during the coldest weather, when the thermometer is 25 to 35 degrees below zero Fahrenheit. The steam dredges have warmed their ponds with the water of condensation and have also used live steam in combating frost and ice. Ice is also chopped from the sides, ladder and chutes. The dredges are manipulated during the cold weather so as to keep the parts most likely to freeze well exposed to the sun. The ground to be dredged is kept flooded and thus does not freeze, and the dredge pond does not freeze if operation is continuous. Amalgamation is not as rapid when the water is cold, but this does not seem to affect the extraction materially. The amalgam is softer and it takes more mercury to do the same amount of work during cold weather.

During the winter of 1907, the coldest in many years, the temperature was continuously near zero during January and at times as much as 30 degrees below zero Fahrenheit. The temperature of the dredge pond was 34 degrees Fahrenheit and the pond was kept from freezing over at times only by moving the dredge about. It was extremely difficult to dig into the corners of the cut by reason of the large accumulations of floating ice.

The ice accumulation on the hull made the No. 3 dredge draw more water and reduced its freeboard at times to about 25 per cent of the normal draught of the dredge. Ice accumulated to such a depth on the ladder that the ladder rollers were out of sight and the buckets slid over ice. Sheaves froze solidly, frozen material accumulated in the hopper, the outboard pump suctions froze if the dredge was shut down for long, and inside small pipes and valves froze and burst. In January about 5 per cent of the total lost time was due to the excessive cold; the dredge operated 72 per cent of the time during the month of January. The Poor Farm dredge has a boiler and steam piping to provide warmth for the dredge men, hot water for clean-ups, and heat for the false-bottomed sluice.

In January 1.7 kilowatt hours of electricity per cubic yard were used. In April only 1.25 kilowatt hours were consumed, but the material excavated was less indurated. The total amount dredged in January, 1907, was 81,000 cubic yards in 539 hours; the total for April was 108,000 cubic yards in 599 hours.

Elsewhere in Montana a 5¾-cubic-foot open-connected Risdon type bucket elevator dredge, electrically driven, is operating at Rocker, near Butte, in Silverbow County. A number of years ago a tail sluice dredge was operating on Gold Creek in Powell County, but whether this dredge is now in operation is not known. No dredging is being done at present on Grasshopper Creek, near Bannock, in Beaverhead County. Considerable prospecting for dredging ground has been done in Montana, but it is understood from men interested at Ruby that so far no successful results have been obtained.

DREDGING AT BRECKENRIDGE, COLORADO.*

Breckenridge, the county seat of Summit County, Colorado, is located 112 miles southwest of Denver, on the South Park branch of the Colorado and Southern railroad. The elevation is 9,600 feet above sea level. The Breckenridge district embraces the valleys of the Blue, Swan, and Snake rivers, together with the drainage basins of their tributaries. These streams head on the western side of the main range of the Rocky Mountains, and consequently drain to the Pacific.

In the present bed of the streams, and along their flood plains, are deposits, varying in width from 100 yards to a mile in places. These are known as "river gravels." They contain many water-worn rocks from 6 to 8 inches in diameter, with occasional boulders as large as 3 feet in diameter. In the Blue and Swan river gravels, which are compact and cemented by clay, porphyry is the predominant rock.

The gravel varies considerably in depth in different places, as it appears to have been deposited on an uneven floor. Near Breckenridge, at the Gold Pan workings, the depth is 76 feet to bedrock; at Dillon, seven miles down the Blue river, it is 79 feet to bedrock; between these points in the Blue and up the Swan rivers the depth is from 40 to 45

---

*Abstract from a paper by A. H. Bradford and Roy P. Curtis, Colorado School of Mines.

feet. The Upper Blue River above Breckenridge shows but little evidence of the influence of waters after their glacial deposition. The deposits are very deep, the boulders large, while the gold is coarse and shows little action of water wear. The large boulders were the prime cause of the failure of an extensive enterprise on the Gold Pan placer deposit. The bedrock of the dredging area consists principally of either porphyry or shale, or shale traversed by dikes of porphyry. The shale is often nearly horizontal. Most of the gold is generally found on or in the oxidized shattered porphyry or on top of the shale, but gold is found from the "grass roots" to bedrock, and sometimes for a few inches or feet into the bedrock.

The pay-streaks occur in channels, not continuous, and the course and dimensions of which can be ascertained only by drilling or prospecting by shafts. The gold is coarse rather than fine, due perhaps to not having traveled far from its source. Large nuggets are uncommon, but small ones half an inch to an inch long and generally flattened are not rare. The fineness of the gold ranges from $17 to $18 per ounce. Black sand occurs in considerable quantity, and is said to have a smelting value of $75 per ton. The black sand is not all magnetite. Much of the gold is coated by iron oxide and therefore does not amalgamate.

While in former years the "bench gravels" have been extensively worked it is the lower river gravels that are now being worked by modern placer mining methods. The early dredges built here were not adapted to the nature of the ground, being too light for the successful handling of the stiff cemented gravels and the boulders. Large quantities of black sand proved a great hindrance, especially on the gold-saving tables, where, owing to its specific gravity, it chocked the riffles and prevented the catching of gold in them. The hydraulic elevator, the clam-shell dredge, and other light constructed dredges have all proved failures, and have been succeeded by the latest and most powerful California type of dredges, which have proved successful in the economical handling of the gravel and efficient saving of the gold.

Between 1897 and 1900 the North American Dredging Company operated three dredges on the Swan River. In 1907 this company reorganized into what is now known as the Colorado Gold Dredging Company, and the construction of two powerful California type dredges at the junction of the Swan and the Blue rivers marked a new era for this company.

The Reliance Gold Mining Company operates a dredge in French Gulch, about a mile and a half from Breckenridge. It is the oldest machine in the district, and has been in successful use for five years. The conditions met with in the immediate vicinity are: (1) Coarse gravel and small rock fragments. (2) Large boulders. (3) Streaks of sticky clay carrying the greater portion of the gold. These streaks must be disintegrated before the gold can be saved. (4) Much of the gold is coarse and rusty and will not amalgamate, and (5) much heavy black iron sand tends to clog the riffles. The Reliance dredge was originally driven by steam and was of the double-lift open-connected bucket type. During the season of 1909 the motive power was changed to electricity, and the dredge remodeled. Before remodeling the dredge extraction was effected by means of a sluice 120 feet long by 4 feet wide, extending behind the stern of the dredge and supported on independent pontoons. The chief objection to this arrangement was that there were no means for stacking the tailing, and the tailing pile soon encroached upon the unworked ground necessitating rehandling. In remodeling, the dredge was equipped with a tailing stacker and the riffle tables, in place of the sluice flume, were placed on board of the main hull, thus putting the gold-saving device wholly on board the boat and allowing the tailing to be stacked far enough behind the dredge so that there is no danger of their impeding the operations of the dredge.

By the addition of a heating appliance for the inside of the dredge and for the exposed parts of the digging ladder the necessity for closing down during the past winter season was avoided, the Reliance dredge now being able to run all the year round. This heating is an important factor for consideration, since the months from November to February are severe. Heat is generated by a 30-horsepower boiler, and as one cord of wood lasts about two days the cost of heating is not excessive.

The electric power for operating the dredge is delivered through an insulated cable from a portable transformer on shore. Alternating current of 13,000 volts is stepped down to 440 at this transformer.

The Colorado Gold Dredging Company has two Bucyrus dredges at Valdoro, five miles down the Blue River from Breckenridge. One is working down stream in the Blue River, the other up stream in the Swan. These dredges have operated but one season. The dredge working up the Swan River is designed to dig 38 feet below the surface of the water and the other 48 feet. The latter dredge has worked down the Swan River and is now at the junction of the Swan and Blue rivers, and will continue down the Blue River. Both dredges are of the open-connected single-lift bucket elevator type. The material is excavated and lifted but once, and progresses by gravity from the time it leaves the bucket until it is discharged back into the pit as waste. There are 42 9½-cubic-foot buckets and 42 links in the chain, and the dredge is said to handle 3,000 yards per day of 24 hours.

The electric power for operating the motors is supplied by the Central Colorado Power Company. The total rated horsepower on each dredge is about 430. The electric power is transformed from 13,000 volts down to 440 volts before being brought on board the boat. The motors operate at 440 volts and the lighting system at 110. The motors used are main drive or digging motor, 200-horsepower; sand pump motor, 75-horse-power; water pump motor, 75-horsepower; stacker, winch, and trommel drive motor, 25-horsepower; stacker motor, 20-horsepower; winch motor, 20-horsepower; and deck pump motor, 15-horsepower.

The hull is 115 feet long, 40 feet 6 inches wide and 9 feet deep. The sides are curved inward at the forward end, making the bow 26 feet wide, and the well is 6 feet 6 inches wide. Both the digging and walking spuds are made of structural steel, with extra heavy cast steel points. The walking spud weighs 24,000 pounds and the digging spud 45,000 pounds.

The Colorado Gold Dredging Company dredged during the last season at a cost of 8 cents per cubic yard, everything included, and as nearly as can be ascertained the average saving was from 12 to 14 cents per yard. The gold saving under the best conditions is estimated to reach 80 per cent.

GOLD DREDGING IN IDAHO--BOISE BASIN.

The Boston-Idaho Gold Dredging Company moved a dredge of 3,000 yards daily capacity from Yreka, California, to their property below Idaho City. A hydro-electric power plant on the south fork of the Payette River, with a rated capacity of about 1,400-horsepower, transmits power about fifteen miles to the dredge.

This company is having a dredge of 9,000 cubic yards daily capacity designed for the same ground, which is quite extensive and is said to contain average values, of 16 cents per cubic yard. The property of the Boston and Idaho Gold Dredging Company belonged formerly to the late Charles Souther of Boston, Mass., who at first installed and operated a dipper dredge, equipped with double hoppers, screens, and stackers. This dredge was not a success. In 1898 a 3¼-cubic-foot open-link-bucket elevator dredge, constructed by the Risdon Iron Works, was installed. This dredge, which was steam driven, operated successfully under the direction of Captain Winters for many years, until worn out. The holdings of the Boston and Idaho Gold Dredging Company comprise an area of about 617 acres, located along Moores Creek and part of Elk Creek, extending from the town of Idaho City, several miles down Moores Creek, along the road to Boise City. Previous to dredging, extensive hydraulic mining operations were carried on along the banks of Moores Creek where the bench gravels were rich in gold. As a result of these operations, Moores and Elk creeks are clogged with hydraulic tailings.

At Elk City, the Ihaho Dredging Company and the Jennings Dredging Company operated but a short time before closing down for the season 1909. The former company has a considerable area of dredgeable gravel on American River; the latter has ground on Little Elk River, three miles north of Elk City.

In the Boise basin the Moline Mining Company has operated a Risdon dredge near Placerville for the past 4½ years, having averaged 8 months per season, 27 days per month, handling 1,400 cubic yards per day, working three 8-hour shifts. This dredge has 5-foot buckets, open-connected, is equipped with a compound condensing steam engine, the fuel for the boilers being wood that costs $3.25 per cord delivered to the dredge, the company owning the timber land from which the wood is taken. An area of about 80 acres of dredgeable ground is owned, the gravel being from 30 to 45 feet deep, and overlying a soft sedimentary bedrock, which is easily cleaned. The operating season for 1909 was slightly shortened by reason of the dredge being dry docked November 1st so as to overhaul and reinforce the hull, an unusual operation that was successfully accomplished, and the dredge put afloat again before the winter freeze up.

*GOLD DREDGING IN ALASKA.

The production of gold from placer mines in Alaska for the year 1909 is estimated at $16,000,000 as compared with $15,885,000 for 1908, and $16,491,000 in 1907. Most of this, with the exception of about half a million dollars, the combined production of the smaller districts, comes from the Yukon Basin and the Seward Peninsula. Information is not available as to what proportion of the placer gold is produced by dredge operations.

Up to four or five years ago, most of the attempts at dredging were either failures or gave but little profit, and the results discouraged even experienced mining men. In spite of adverse opinions and the failures of the first attempts, many engineers persisted in their experiments to meet the local conditions and it is owing to the efforts of these men that dredging is being successfully carried on in Alaska and adjacent parts of Canada.

The installation of two additional dredges near Nome and the successful operation of those of the Yukon Gold Company, several in the Seward Peninsula and three in the Forty-mile district, one of the most isolated placer districts in Alaska, are significant of the future of Alaska placer operations and indicate that progress is steadily being made to work the lower grade gravels.

In the Seward Peninsula are considerable areas of gold-bearing gravels. In 1909 there were about eleven dredges working, including several that were installed the latter end of the season, and a number of dredges have been ordered for the spring of 1910.

---

*Partly abstracted from "U. S. Geological Survey Records" and the "Mining Press."

The dredge of the Three Friends Mining Company, Solomon River, Seward Peninsula, operated from the middle of June to the middle of October, which is the full dredging season in that district. It is a 5-cubic-foot close-connected-bucket elevator dredge, California type, constructed by the Western Engineering and Construction Company of San Francisco, and equipped with Bucyrus Company machinery. It was erected complete in the year 1905, in the record time of 57 days, and has been operating continuously ever since.

Up to the close of the season of 1909, the Three Friends dredge had dug 92 acres of ground, the average per day being 3,706 cubic yards, or 111,180 cubic yards per month. This is considered an exceptionally

No. 230.   Three Friends' Dredge, Alaska.   California type.

good record for a dredge operating under the severe conditions met with in Alaska. In 1908 it was necessary to dig 6 feet into a hard limestone bedrock, in order to extract the gold, a condition which existed for over four weeks. Aside from excessive wearing away of the lips, the operations of the dredge were not affected, a fact which reflects credit on both the management and the builders.

Steam power is used, coal being shipped from the outside at a cost of $20 a ton at the boat. The actual operating costs are said to be 11 cents per cubic yard, and the total costs including winter care, repairs, and amortization of capital are said to be 18 cents per cubic yard.

The Nome, Montana, New Mexico Mining Company, also on Solomon River, operates a 5-cubic-foot open-connected bucket dredge, built by the Risdon Iron Works. This dredge uses 4.6 tons of coal per day for fuel and is said to average 60,000 cubic yards per month. For some months the operating costs were 11 cents per cubic yard; the ground averages 9 feet in depth. The plant is equipped with compound condensing engine, surface condenser, and locomotive type of boiler, and is considered one of the most economical operated in the district. This dredge had a season of 143 days in 1909.

A dredge that would be regarded by most dredge operators as a toy, but which, nevertheless, is said to have made a profit of $15,000 for the season is the Sievertson Dredge, also on Solomon River. It is a small machine driven by gasoline engines and equipped with 1-cubic-foot open-connected buckets; it is reported to have cost $6,000. A number of

No. 231.   Timber construction of dredge hull, Alaska, as designed by the Western Engineering and Construction Company.   California type.

other companies are operating dredges, but results are not available at this writing.

There is undoubtedly a field in the Seward Peninsula for dredges with 3- and 5-cubic-foot buckets and several installations are to be made this season. The stream beds are mostly shallow, being from 10 to 25 feet deep, the bedrock is generally schist or limestone and the gravels loose and generally free from boulders and clay.

The record of the dredges of the Yukon Gold Company for the past season is noteworthy, especially as the ground had to be thawed by steam. The Yukon Gold Company operated seven dredges in 1909. They commenced operations as early as power was available. The last dredge to start began June 9th. The dredging season for six of the seven dredges was 132½ days. The running time of dredge No. 5 was curtailed on account of local conditions. During the season the dredges handled 2,381,800 cubic yards and produced $1,363,722 worth of gold. The value per cubic yard was 57.24 cents and the cost 31.94 cents per cubic yard. This cost included thawing charges, amounting to 15.45

No. 232. Rothchild No. 1 Dredge operating in Alaska. Note electric power plant and wood in distance. Marion make. California type.

cents, preliminary stripping operations, and depreciation at the rate of $2,000 per month per dredge.

A remarkable record was made at the power plant at the head of Little Twelve-mile River, the loss of time at the power house was only twenty minutes in the entire season. The capacity of the plant is 2,000 horsepower.

It is reported that the Detroit Gold Mining Company, the Lewis River Gold Mining Company, and the Walkers Fork Gold Dredging Company, each operating one dredge, had a successful season in 1909.

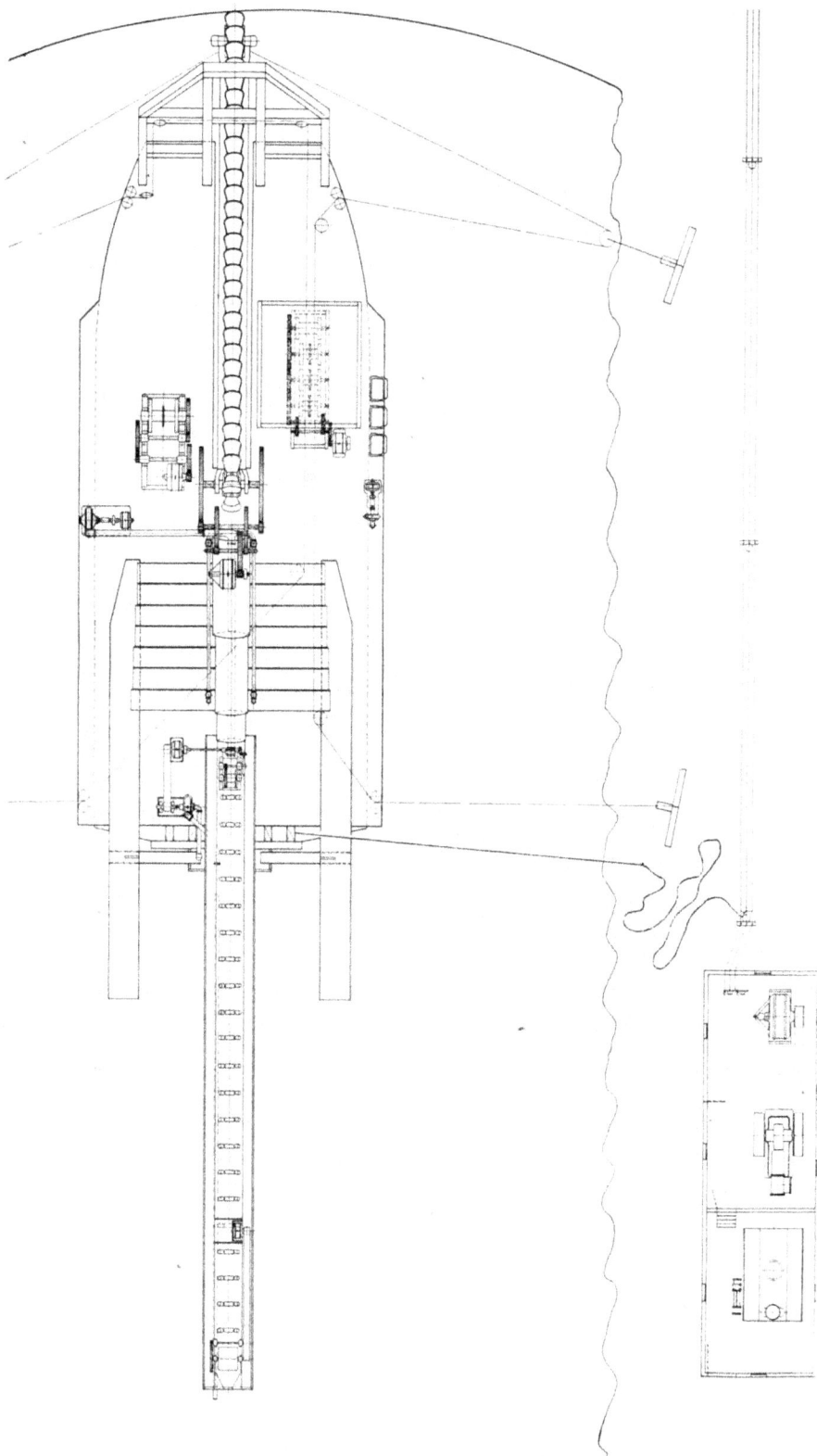

No. 233. Plan view of electric driven placer mining elevator dredge with steam plant on shore. Marion Steam Shovel Company design. California type.

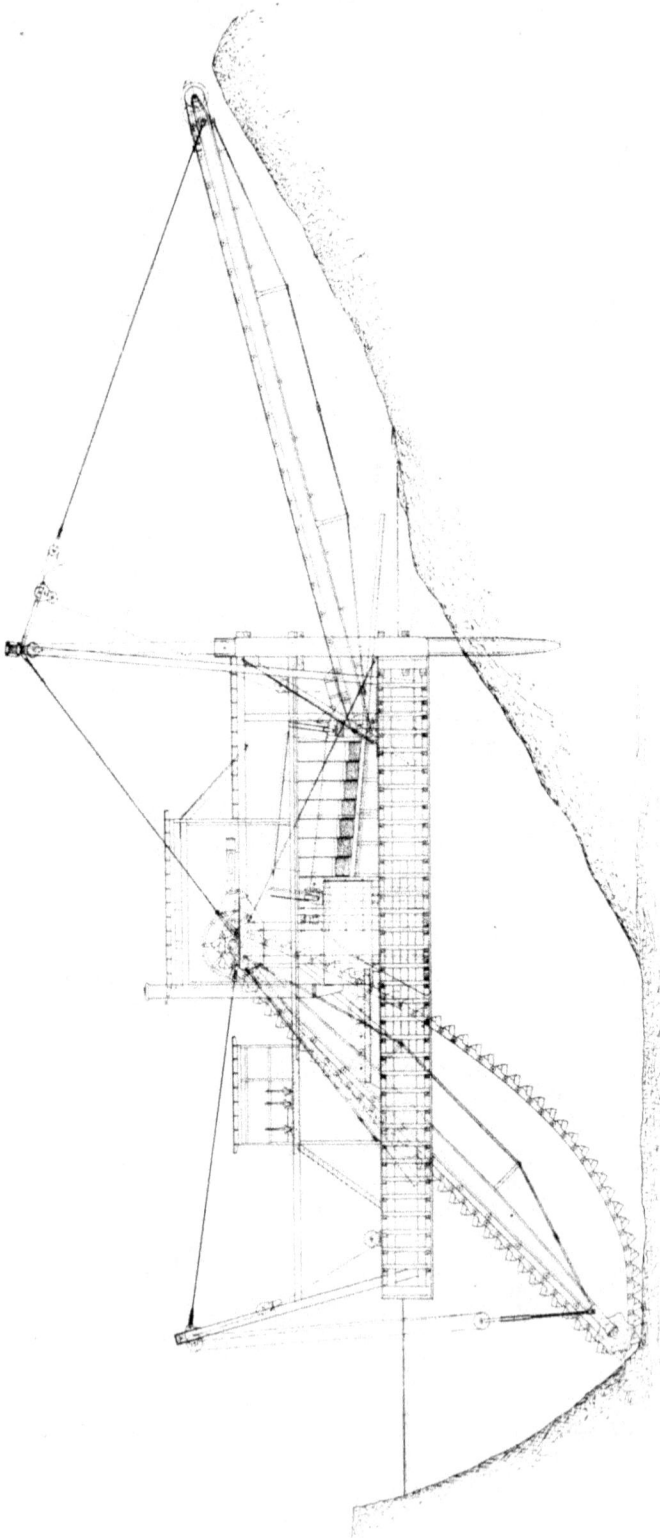

No. 234. Side elevation of steam driven placer mining elevator dredge. Marion Steam Shovel Company design.

The following information is given by Joseph W. Boyle, manager of Canadian Klondyke Mining Company, Limited, operating on Bear Creek, near Dawson, Alaska:

The Bear Creek dredge, a 7½-cubic-foot bucket California type elevator dredge, electrically driven, constructed by the Marion Steam Shovel Company, started dredging this year (1909) at 11 p. m. on May 9th and stopped at 7 a. m. November 20th, a total of 194 days and 8 hours for the season. The boat dredged 681,616 cubic yards of material, or on an average of 3,558 cubic yards per day throughout the season. The total lost time for the season was 368 hours and 43 minutes, which was 7.90 per cent of the total possible running time, which indicates that the length of season is greater than was supposed possible in Alaska. It is expected that the dredge will operate 224 days next season.

The operation of this dredge for the above period compares favorably with those of the best boats in California, both as to the yardage handled and the lost time during the operating season. Mr. Boyle states that the actual operation of the dredge and power plant, including maintenance and repairs and all items of labor, fuel, etc., has for the year 1909 been less than 15 cents per cubic yard of material handled, which he hopes to reduce next year.

Heating appliances were arranged with steam pipe extending on both sides of the entire length of the enclosed stacker ladder, with steam coils under the lower belt rollers, under the sluice tables, and in the enclosed house constructed over the main drive and all other points necessary to be kept free from frost. On different occasions during the month of November, 1909, the dredge was operated in a temperature of more than 20 degrees below zero without the slightest difficulty and with all the efficiency which it would have in the warmest weather. The system of applying hot water to the fair lead sheaves through which the various shore lines were operated prevents any difficulty from this quarter.

A number of dredge installations have been failures in Alaska, as elsewhere, principally because they were based upon ill-advised schemes or because of the lack of proper technical supervision. As a rule, each property should be studied carefully, by a competent engineer, to determine if the proposition would justify exploitation and the best method of working.

One of the heaviest items of expense in Alaskan placer mining operations is that of transportation, which affects not only the initial installation, but also operating expenses, including labor and fuel. The present price for labor varies from $5 to $6 a day and board, making a total cost per man of $6 to $8.50 per day in the most important camps.

Coal costs from $15 to $18 a ton at Nome; to this must be added the cost of transportation to the scene of mining operations. Where wood

No. 235. Placer mining plant in the Yukon. See page 274.

is available, the prices run from $7 to $12 a cord, delivered to the mines. One of the many factors that affect dredging is the frozen character of the ground which has been described in various articles on Alaska. A few years ago it was considered impossible to profitably dredge frozen ground, but experience has shown that this is not always the case.

PLACER MINING IN THE YUKON.

On Bonanza Creek the ground being worked by hydraulic giants and sluices is 20 to 25 feet deep and very flat. To obtain the necessary dump a portable bucket elevator was arranged to raise the gold-bearing gravel from the bedrock sluice to a suitable height and discharge into a system of portable sluices arranged to facilitate a change of position when necessary to obtain a new dump. The whole of the elevator machinery is supported upon a structural frame mounted on wheels, one day being required to move the machine to a new position and put it in order ready for operation.

This method has proven very satisfactory, and at the present time O. B. Perry, general manager and chief engineer for the Yukon Gold Company, has three of these machines of his own design at work upon the company's property. The mode of operation is as follows:

A sump approximately 20 feet square, having a depth of 14 to 16 feet below bedrock, is excavated to receive the lower end of the elevator, a channel or bedrock sluice emptying into the sump, having an approximate grade of 5 inches in 12 feet, is excavated in the bedrock and provided with riffles. The bank of gravel to be treated is washed down by two hydraulic giants with 3-inch nozzles; a third giant is directed down the bedrock sluice to carry the gravel to the elevator which dumps the gravel into a riffled sluice, approximately 25 feet above the ground.

The elevator chain consists of 3-cubic-foot close-connected buckets, operated by a 50 horsepower motor. The water used in the upper sluice is pumped from the sumps by one 12-inch centrifugal pump belted to a 100 horsepower motor and one 8-inch centrifugal pump belted to a 50 horsepower motor.

A derrick having a long boom is located in a convenient position to handle the large boulders and deposit same in a suitable place clear of the gravel being treated. See illustration, No. 235, page 273.

*GOLD DREDGING IN THE PHILIPPINE ISLANDS.

The only successful dredging in the islands has been carried on in the Paracale district, which is situated in the northern part of the province of Ambos Camarines on the eastern coast of the Island of Luzon. It is reached by steamer from Manila around the southern end of Luzon to Mercedes, about 20 miles east of Paracale.

*Abstract from "The Mineral Resources of the Philippine Islands."

The Paracale district was regarded by the Spaniards as the most important mining region in the islands, and the mines have been worked intermittently and the streams washed since before the days of the Spanish conquest and the hills are honeycombed with ancient workings.

Since the American occupation development has been hindered by the confusion which has existed in regard to titles, there being great uncertainty as to the validity and extent of many of the Spanish concessions. Lately, however, satisfactory arrangements have been made in many cases and work is progressing.

The Paracale district consists of maturely eroded hills of metamorphic and igneous rocks containing quartz veins. The large plains of the Paracale and Malagit rivers, as well as smaller valleys, afford very promising dredging ground, which is now being thoroughly prospected, chiefly by means of small drills, and showing excellent returns. The Paracale Dredging Company has had a dredge in operation on its property near the town of Paracale for the last year and a half, and such excellent results have been obtained that several other dredges will be in operation in a short time on neighboring properties. The placer ground in the vicinity of Paracale generally consists of about four to five meters of barren clay mixed with vegetable matter. In places this overlies a few centimeters of coral and below this is a varying amount of gray clay carrying values. Beneath this again is an irregular amount of extremely rich sand and quartz pebbles, the latter often showing large amounts of free gold. The gold brought up by the dredge now working is remarkably angular and often shows distinct crystalline structure. The quartz pebbles are often sharp and angular, showing that they have traveled but a very short distance. The bedrock appears to be a schistose rock, decomposed to a clay, which is easily cut by the dredge buckets, making it possible to secure practically all of this rich gravel.

The dredge at present in operation is of the New Zealand type and differs from those generally used in America in that it has no stacking ladder and no quicksilver is used in the riffles. According to the dredge-master's reports, during a period from May 25 to December 31, 1908, 50,244 cubic yards were handled and 2,814.1 ounces of gold, having a value of $50,653.80, recovered. For the whole period from May 25, 1908, to July 1, 1909, a return of 4,985 ounces, valued at $89,731, United States currency, is reported. The dredge is equipped with one 70-horsepower boiler, 60-horsepower engine, a digging ladder 62 feet in length, a revolving screen, and 37 digging buckets, each of 4½-cubic-foot capacity. The hull is built of Oregon pine and is 114 feet long, 30 feet wide, 5 feet 6 inches deep at the bow, and 9 feet deep at the stern. The greater depth at the stern is required to support four sluices, 80 feet long, which have an overhang of 30 feet beyond the stern of the

[Mineral Resources, 1908.]

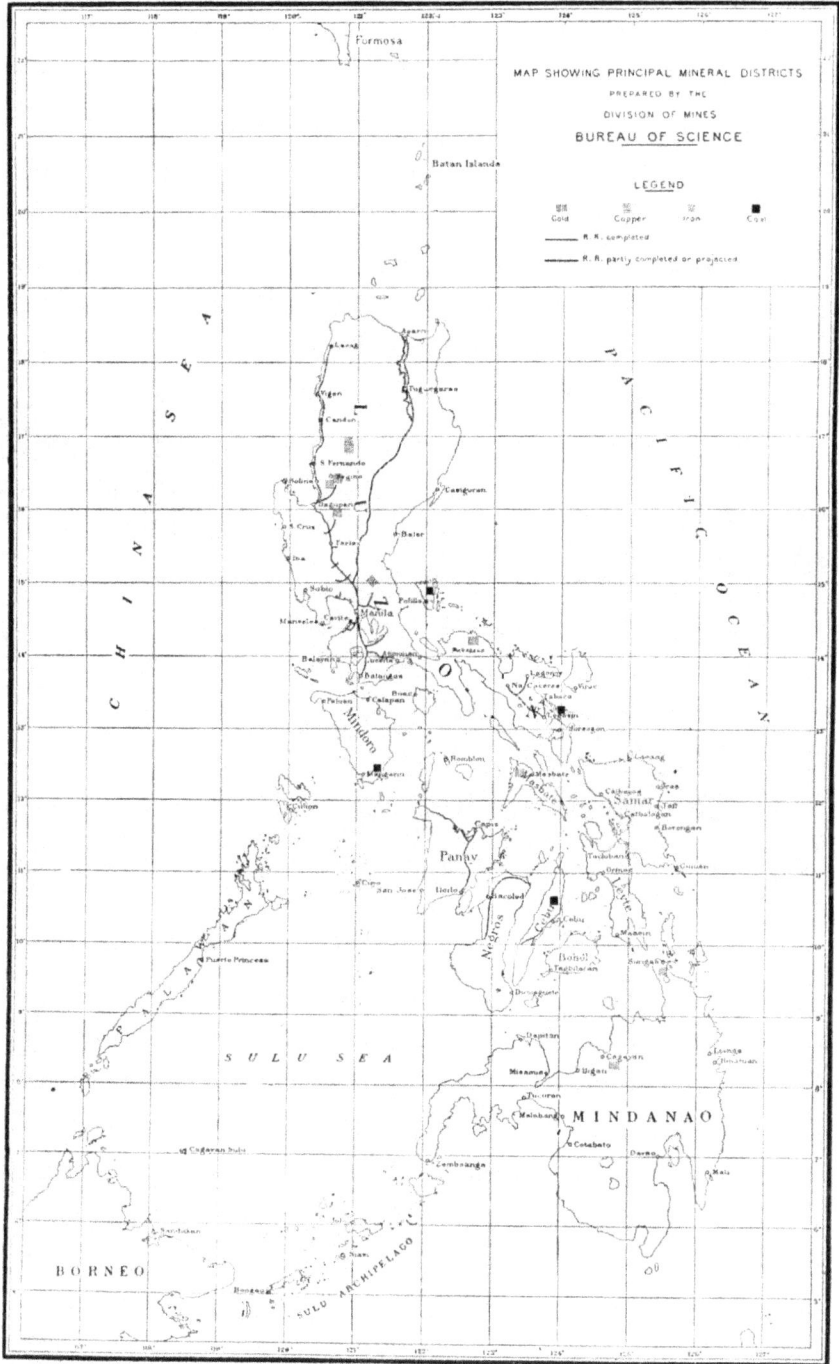

No. 236.   Map of Philippine Islands.

boat. The general system of the gold-saving arrangement includes a big spread of expanded metal and cocoa matting. The custom in the district is to have five white men on a dredge; the dredgemaster, three winch-men, and one extra man, who should be an engineer and machinist. All the other labor is native. The dredgemasters receive $200 to $250 per month and board, and the white men $125 to $150 and board. The natives receive 37½ to 50 cents per day, according to their ability.

Dredging is made difficult not only by the large amount of barren clay which must be removed, but the unusually large percentage of fine

No. 237. Dredge on the Paracale River, Philippine Islands. New Zealand type.

material handled, about 80 per cent of all the material passing through the screens. The black sand carries values and is saved for future treatment, as is also part of the gray clay. The oversize material from the revolving screen is sorted by hand and the quartz pebbles saved for future crushing.

The operations referred to have been carried on by a New Zealand corporation, the Paracale Gold Dredging Company, Limited, working on shares with the claim owners. A second boat has been installed farther up on the same river by another New Zealand company, the Stanley-Paracale Gold Dredging Company. It differs from the first dredge in that it has no revolving screen or tables, the buckets discharg-ing directly into a larger sluice box. A third dredge installed on the Paracale River is a Risdon dredge equipped with 3¼-foot buckets. This dredge was originally set up on the island of Masbate. It was bought by the present owner, the Philippine Gold Dredging Company, and removed to Paracale.

The Paracale River is really an arm of the sea with a rise and fall of some five or six feet of tide. The flat in which it runs is perhaps a mile in width near the mouth. It continues about the same width for a mile or more up and then spreads out into smaller flats with ranges of hills cutting in between. The total area available for gold dredging in this flat is estimated at about 1,300 acres. The average depth to bedrock on the lower river is from 30 to 40 feet. Where the Stanley dredge is at work at a point higher up on the river, the depth is considerably less. A great deal of sand and other fine material is found and probably not over 35 per cent of the dirt handled can be stacked, and of the three dredges only the Risdon has a stacker.

The Malagit River, which flows into the Pacific Ocean not far from the Paracale River, has been prospected, to a considerable extent with hand drills. There is a much larger percentage of gravel and heavy material there than on the Paracale and several other streams in the same district, all tidal rivers running into the Pacific Ocean, are being investigated at the present time.

The Labo and Malagit are the principal rivers, and with their tributaries drain most of the country, which, with the exception of Mount Bagacay in the southeastern portion of the district, about 3,000 feet high, and May Cruz Mountain, forming Mambulao Point in the northwestern part of the district, about 1,000 feet high, is quite flat. Low hills and ridges divide the tide water streams, and except where cleared by the natives for growing hemp, are covered with heavy timber and a dense undergrowth so that prospecting is difficult. The country is inhabited mainly by Tagalogs, though there are a few Bicols left and Negrites from the interior are occasionally found on the small farms. The rainy season is opposite to that of Manila, the heaviest precipitation occuring in December, January, and February.

During the year 1909 there are several improvements planned for the district aside from the mining work. The most important of these will be the removal of a reef which obstructs the entrance to the Malagit River, thereby affording excellent harbor facilities. At present there is no shelter nearer than Mercedes on the one hand and Mambulao on the other, and during bad weather the regular steamers can go no farther than Mercedes, some 30 kilometers down the coast.

Other improvements planned for the district are the building of a telephone line and the construction of a wagon road from Indang. The association also hopes to secure from the insular government some arrangement which will permit of having a resident mining recorder. At present there is unnecessarily great delay and confusion caused by the necessity of sending all papers to Nueva Caceres, the capital of the province.

DREDGING IN NEW ZEALAND FOR THE YEAR 1908.

During the past year, as will be seen from the following statement, the average yield per working dredge was £3,039, being a decline of £239 on that obtained during the previous year; a decline of £45,816 has to be recorded in the value of gold production by dredges, and the number of working dredges has decreased by five.

The following statement shows the number of dredges, the gold produced by them, and the number of persons employed during 1907 and 1908:

	Number of Dredges.				Number of Persons Ordinarily Employed.	Yield During 1908.	Average Yield per Dredge 1908.
	1907.		1908.				
	Idle.	At Work.	Idle.	At Work.			
West Coast	2	25	4	23	196	£75,670	£3,290
Southern	35	103	13	100	817	298,148	2,981
Totals	37	128	17	123	1,013	£273,818	£3,039

The dividends paid by forty-one of these dredges, the property of registered companies, amounted to £75,050 during the year; the profits of privately owned dredges is unobtainable.

The following is a statement regarding the most productive dredges during the year:

Name of Dredge.	Production During 1908 of all Dredges.	Dividends Paid by Forty-three Dredges Owned by Registered Companies.	
		During 1908.	To December 31, 1908.
West Coast District—			
No Town Creek	£7,174	£3,000	£23,400
Pactolus (two dredges)	14,391	8,125	56,250
Southern District—			
Alexandra Eureka	5,012	2,400	18,450
Golden Treasure	5,847	2,445	23,439
Koputai	6,952	2,538	4,813
Manuherikia	5,145	2,100	30,600
Masterton	9,206	5,500	25,000
Mystery Flat	9,873	6,614	20,119
Otago (3 dredges)	5,621	1,250	16,000
Paterson's Freehold	7,669	3,600	13,200
Rise and Shine	14,414	5,700	15,600
Rising Sun	8,567	3,600	8,000
Waikaia	8,113	4,550	10,500
Waikaka Syndicate	5,735	1,400	13,300
Waikaka United (two dredges)	11,767	7,000	33,600
Waikaka Queen	4,345	1,590	4,628
Other working dredges (both districts)	243,987	14,408	377,007
Totals	£373,818	£75,820	£693,906

In Otago, the principal gold dredging field in Australasia, many of the claims on the River Molyneux and upon the Waipori and Waikaka fields are gradually becoming worked out, but upon the upper Clutha and elsewhere considerable areas remain to be worked.

The era of the dredge of small dimensions and low power is past, but there yet remain several years of prosperity for those large dredges of greater power capable of working increased quantities of low-grade gravel.

In Otago and Southland eight dredges are now operated by hydraulic power and two by electricity. The most consistently remunerative dredging field is still that at Waikaia, and some excellent returns have been obtained from the Waikaia, Koputai, Mystery Flat, and Masterton dredges. The first named established a record for land dredging in New Zealand during one week in April, 1908, by securing 248 ounces of gold. The Hartley and Riley dredge in the Clutha Gorge in less than three weeks during the same year obtained from a small patch 450 ounces of gold.

On the West Coast no new dredges have been built during the year, but those at work have done very well, especially the Pactolus and No Town Creek dredges.

The diminution of the dredging returns during 1908 may to a certain extent be attributed to the flooded state of the rivers during the autumn and early winter, when, owing to the height of the water, the dredges were precluded from working in the river gorges for a considerable period.

The following is a statement comparing the production of dividends paid by and number of men employed upon all the gold dredges in Victoria and New South Wales with those in New Zealand, the property of registered companies, during 1908:

	Number of Dredges	Value of Bullion.		Dividends.		Men Employed.	
		Total.	Per Dredge.	Total.	Per Dredge	Total.	Per Dredge.
Victoria (all dredges)_____	47	£220,632*	£4,694	£59,249	£1,260	641	13.6
New South Wales (all dredges)_	26	155,770	5,980	†	†	†	†
New Zealand (dredges the property of registered companies only) _____	41	353,104	8,612	75,050	1,830	369	9

* Estimated at £4 per ounce.
† Not known.

NOTE.—Abstract from report of the Department of Mines on the gold fields of New Zealand for the year 1908.

GOLD DREDGING IN SIBERIA.

The gold mines of Siberia proper consist almost entirely of river placers. The Ural region has a number of quartz mines and about two fifths of its gold production comes from this source. The gold-bearing gravels of Siberia are widely distributed, each of the great rivers having among its tributaries several areas from which placer gold is produced.

The richest placers are probably those of the Lena region, where $25 to $30 per cubic yard is not uncommon. In the valley of the Yenisei, along many of the tributaries of the Amur, and elsewhere throughout Siberia are placers that would cause big stampedes if located in America or in English territory.

Siberia is by no means an unknown country, and is far from being a frozen land. A report by Reutofski, a geologist in the government service, gives much that is of interest and is accompanied by geological maps. Unfortunately, this work is in Russian and has not yet been translated to English. Numerous articles have been published in the "technical press" of this country and England on Siberia, abstracts from which have been freely taken when available in writing this article, and also from Reutofski's report and the statistics furnished by the Gold and Platinum Miners' Association of Russia.

According to Reutofski the Siberian placers have yielded, up to 1904, at least $700,000,000; the Nakatami River, in the Lena district, yielded between 1865 and 1900 $54,000,000 from gravels, having an average value of $10 per cubic yard. On the Akakan, one of its tributaries, the pay streak is reported to be from 300 to 500 feet wide and has been worked for three miles along the stream; the gravel is $3\frac{1}{2}$ to 15 feet thick and carries a heavy overburden.

Gold dredges are of recent date in Russia and were first installed about seven or eight years ago. Gold mining has existed in Russia for over 150 years. On account of the crude methods employed in washing auriferous gravel the tailing of old plants should present good opportunities for the work of dredges. An experiment to work with modern methods the tailing from an old placer mine was successfully carried out by Mr. C. W. Purington in the Amur district, over 60,000 cubic yards being handled in the test and a total clean-up of $67,000 obtained.

There have been many failures recorded of dredging operations in Siberia, but it is not surprising considering the faulty equipment of most of the early installations. Many of the first dredges were of German make, built by people who had little or no experience in gold dredging, and were often installed on a proposition under terms whereby the construction company received the gross profits until the dredge was paid for. The gold-saving apparatus was crude and the machinery was not durable, but generally held together long enough for the Germans to get their money from the gross profits.

With the installation of modern dredges, notably those built by the Putiloff Works at St. Petersburg, from plans of California type dredges furnished by the Bucyrus Company, a new impetus was given to the dredging industry in that country. Details of operation of two California type dredges built by the Putiloff Works and installed and successfully operated at Ivdell are given by W. H. Shockely in the *Mining Magazine*.

Details of Ivdell Dredges.

Size of Buckets.	Worked for		Yards Exca-vated.	Ounces of Gold Pro-duced.	Yield per Cubic Yard in Cents.
	Days.	Hours.			
No. 1, 3-foot	192	3,783	156,210	1,655	18.8
No. 2, 5½-foot	184	3,283	256,159	1,962	13.4

NOTE.—Work began April 14, 1907, and ended October 19, 1908.

The Permanent Consultive Office of the gold and platinum industries in Russia, which represents the interests of the gold and platinum industry of all Russia, has published interesting information on the production of the Ural and Siberian dredges in Russia for the years 1906, 1907, and 1908.

The average number of days worked by dredges reporting the season of 1908 was 149½ days, one of the dredges of the Neviansk Company at Nikolaievsk operated 194 days. Most of the dredges commenced work late in April and in the Yenesie region about the middle of May. Some of the dredges were shut down by the first of October, but the majority continued digging until nearly the end of October and a few well into November. Two dredges in the Nijni Tagil region of the Urals are reported as working up to the 11th of December. The information furnished by the 49 dredges reporting out of some 64 dredges known to be in existence in Siberia and the Ural region in 1908 is not all that could be desired, but serves to give an idea of dredge operations in that country.

Information is not included in the report concerning the type of dredges used. They varied greatly in capacity and size of buckets; twenty having 7-cubic-foot buckets and most of the balance 5 and 4½-cubic-foot buckets. The average yardage handled during the season was 166,400 cubic yards per dredge. The following table gives average returns and yardage handled by reporting dredges for 1906, 1907, and 1908:

	Number of Dredges at Work	Reporting Dredges	Average Working Days	Working Time Average	Cubic Yards Handled per Dredge	Ounces of Bullion Recovered per Dredge	Approximate Recovery per Dredge	Approximate Value per Cubic Yard in Cents
1906	40	32	173	2,837	159,600	1,082	$19,205	12
1907	64	46	137¾	2,259	139,300	1,140	20,235	14.5
1908	64	49	149¼	2,502	166,400	1,430	25,380	15.2

For the year 1908 the average recovery per dredge was 1,430 ounces, or, approximately, $25,380 for the season. The average value of the gravel worked by all the dredges, according to the reports, was a little over 15 cents per cubic yard, one dredge reporting the gravel handled at a little over 31 cents per cubic yard.

The bureau estimates that the average cost of carrying on dredging operations in the Ural region was about $15,000 per dredge, exclusive of amortization, while $19,000 is figured as the expense per dredge for those operating in the Yenesei region, where conditions are less favorable. Taking $5,000 as the average amount to be set aside annually to amortize the cost of the dredge and equipment, it is estimated by the bureau that the operations in the Ural show an average profit of $5,000 per dredge, while in the Yenesei region they barely pay expenses.

These average figures are of little value on account of the various conditions of operation. Many dredges worked under conditions that made it impossible to more than roughly estimate the quantity of material lifted.

Siberia offers many inducements for gold dredging operations and is attracting much attention from capitalists and investors. A number of California engineers have reported on different fields and some are at present engaged on such work, mostly for English companies.

The following interesting description of the conditions met with in Siberia has been kindly furnished by Mr. J. B. Landfield:

A foreign company has no difficulty legally in operating in Russia, provided the laws are observed. The one definite requirement is that the company have a responsible representative residing in Russia upon whom can be served any papers or processes. Some corporations follow the practice so much used in connection with Mexican mines, i. e., they form a Russian corporation and then have a foreign holding company. In addition to the strictly legal conditions there are important practical conditions that affect the successful operation of a foreign corporation in Russia. The most important is the question of language. The use of an interpreter is so unsatisfactory that the corporation that has not a Russian speaking manager, or one that will quickly learn the language and the mining law which, in general, is an excellent code and far superior to our own, had better not attempt to operate.

Labor is good if properly handled. The Siberian Russian is strong and hardy. In southern Siberia the Tartars make excellent workmen. The Chinese and Koreans in eastern Siberia in many cases are splendid workmen, and are quiet and law-abiding, but there is a strong exclusion sentiment in Russia, and it is likely that they will have to go, or that restrictions will be placed on their employment.

There are many placers far removed from easy transport, which point must be taken into consideration; but the railroad and the numerous rivers reach a great many regions, and the snow in winter makes possible the cheap hauling of heavy machinery on sledges. In many cases in order to take advantage of the winter roads it will be found necessary to allow an extra season before the dredge can be made ready for operation. Supplies of food, fodder and simple necessaries are usually cheap, much more so than in corresponding Alaska camps. General supplies can usually be easily and cheaply transported by taking advantage of the proper season to lay in stores.

In regard to labor it must be said that the Russian workmen is remarkably quick to learn the handling of machinery. The majority of them seem to be natural mechanics, and it is easy to train up efficient help for the mechanical end. Superintendents and assistants are difficult to obtain in Russia; that is to say, efficient ones. The technical education in Russia is to blame for this. The course is largely theoretical, and the graduate is too proud to work with his hands. Furthermore, there is seldom developed any sense of personal responsibility.

In general, it may be said that the conditions of labor and transportation facilities ought not to affect operation and maintenance greatly, but the latter will, in most cases, raise the original cost of the plant.

Most Russian placers are rather extensively prospected according to the methods employed there; that is to say, a Russian mine owner spends a great deal of money on shaft sinking. These shafts or pits are put down in rows and the pits are usually 35 feet apart (sometimes 70 feet) and the rows from 140 to 350 feet apart. A post near each is marked with the number of the row, the number of the pit, and sometimes the date. The superintendent has a book specially ruled for keeping the records of the prospecting. Each shaft has a page and one line is devoted to the record of each seven inches of depth. The reason for this is that the arshin, the regular Russian measure, is 28 inches and the quarter arshin, also a familiar measure, is found convenient for units of material to be washed. There are columns marked "overburden" and "pay-streak" and under one or the other column will be entered such remarks as "fine gravel," "blue clay," and so forth, for each seven inches. Tests are made from the gravel of each seven inches of depth and these tests mean the washing of a considerable amount of gravel. The gravel is washed in a washing machine of the regular Siberian type, only smaller, and consists of a long tom, followed by a short, steep sluice. As most prospecting is done in the winter time the gravel is carried to a cabin for testing. In spite of all this work and apparent care, however, the results are very inaccurate. There are two reasons for this, the carelessness in testing, owing to the crude machine, and the lack of a feeling of personal responsibility on the part of the superintendent.

The placers of Siberia, scattered over an area of 5,000,000 square miles, present every variety of depth, bedrock, etc. In general, the river placers of western, southern, and eastern Siberia are not very deep. Except in some instances in the lower reaches of the rivers they are seldom more than 15 to 18 feet in depth. In the Lena region, however, they are sometimes of great depth, buried under the drift. Fine bedrock, from a dredging standpoint, is frequently met with. Many placers have a bedrock of decomposed diorite or other rock soft enough to be taken up by a dredge to a depth of 18 inches.

All sizes of boulders are met with, but in some placers there are none of any size. In most localities the ground freezes in winter and this delays work in the spring. Only in the extreme north is permanently frozen ground met with. The sticky clay that frequently lies on top of the gravel in the upper stretches of a river placer is a serious obstacle to dredging. In washing it rolls up into balls and interferes with the saving of the gold, and it is often greasy enough to make the water carry off any gold of a flaky character. In going down the streams this clay usually disappears. One curious feature of Siberian placers is that the trees have very shallow roots and most of them can be easily overturned.

The character of the gold varies greatly. In some placers the gold is very fine, almost like powder, and yet so heavy that it was easy to save practically all of it. In other placers the size is very irregular. In one locality there was so much silver in the gold that it was very light and hard to save in the upper reaches of the stream, but attained a much higher fineness lower down. Gold in Siberia seldom fails to amalgamate readily. There are as great variations in fineness as in the gold in different parts of Alaska.

Wood is the chief fuel in the placer regions of Siberia. Except in the Altai region, and perhaps the Maritime province and the Ural region, there seems little chance for the development of water power. Coal and oil are not found in the neighborhood of the placers, and transportation on them is prohibitive. Wood is usually plentiful and costs about $1.50 per cord. There is very little market for electric power in the mining localities.

GOLD DREDGING IN MEXICO.

Extensive "dry" placers occur in the Altar district near La Cienaga and Palomas, in the State of Sonora. Numerous attempts to work these placers have been made with varying success and those near Palomas, it is reported, will be worked by dredges in the near future. Other areas of auriferous gravel occur in southern Sonora and in Sinaloa. An attempt was made to dredge gravel near Suaqui Grande, Sonora, some years ago, but the venture proved a failure.

GOLD DREDGING IN COLOMBIA.

The greater part of the gold yield of Colombia has been from placer deposits. All the large rivers passing through Antioqua are auriferous, as also are or have been the majority of the smaller tributaries. Of the former the principal are the Magdalena, Cauco Porce, and Niehi. At Zaragoza, the placers were discovered in 1581, and are said to have yielded over $5,000,000. Hydraulicing has been carried on successfully at various times on the Porce and Niehi rivers, but all attempts at dredging, up to the present time, have been failures.

Promising placer deposits in Colombia are those of the Choco district, along the Atrato and San Juan rivers. The unhealthy climate and difficulties of transport have hitherto prevented the successful exploration of these placers by modern methods, though several hydraulic claims have been worked near the heads of the rivers with, it is claimed, good results.

The most important known dredging field in Colombia is on the Nichi River, near the town of Zaragoza, where an area of 300 acres has been proven payable dredgeable ground. The gravel is said to average 11 yards deep and the value to be 30 cents a cubic yard. This property is controlled by the Oroville Dredging, Limited, and is part of a concession of 22,000 acres, the greater portion of which has not yet been prospected.

A hydro-electric plant is being installed and a California type 8½-cubic-foot-bucket elevator dredge, with steel hull, has been designed by the Yuba Construction Company, Marysville, California, and will soon be shipped to the property. The following abstract from an article by A. P. Rodgers in *The Engineering and Mining Journal,* gives an idea of the general conditions in the country.

Colombia offers to-day several attractive features for gold dredging operations, although the drawbacks should be well understood before any one attempts to invest his money there. I spent several months upon a thorough examination, which resulted favorably, of a piece of land in that country and as far as I know it is the first expedition ever sent down there that has drilled the ground in the systematic manner that such work is done in California and other established fields. The successful operation of a dredge upon this land will be watched with much interest, and it is bound to prove a stimulus to other similar projects within a short time, when the work of equipment has been completed and the extraction of the gold has actually begun.

The conditions under which one must work are radically different from those found in the United States. This land is all situated within 10 degrees of the equator, and most of it is at an elevation above sea level, which is considered unfavorable to health in that latitude. The

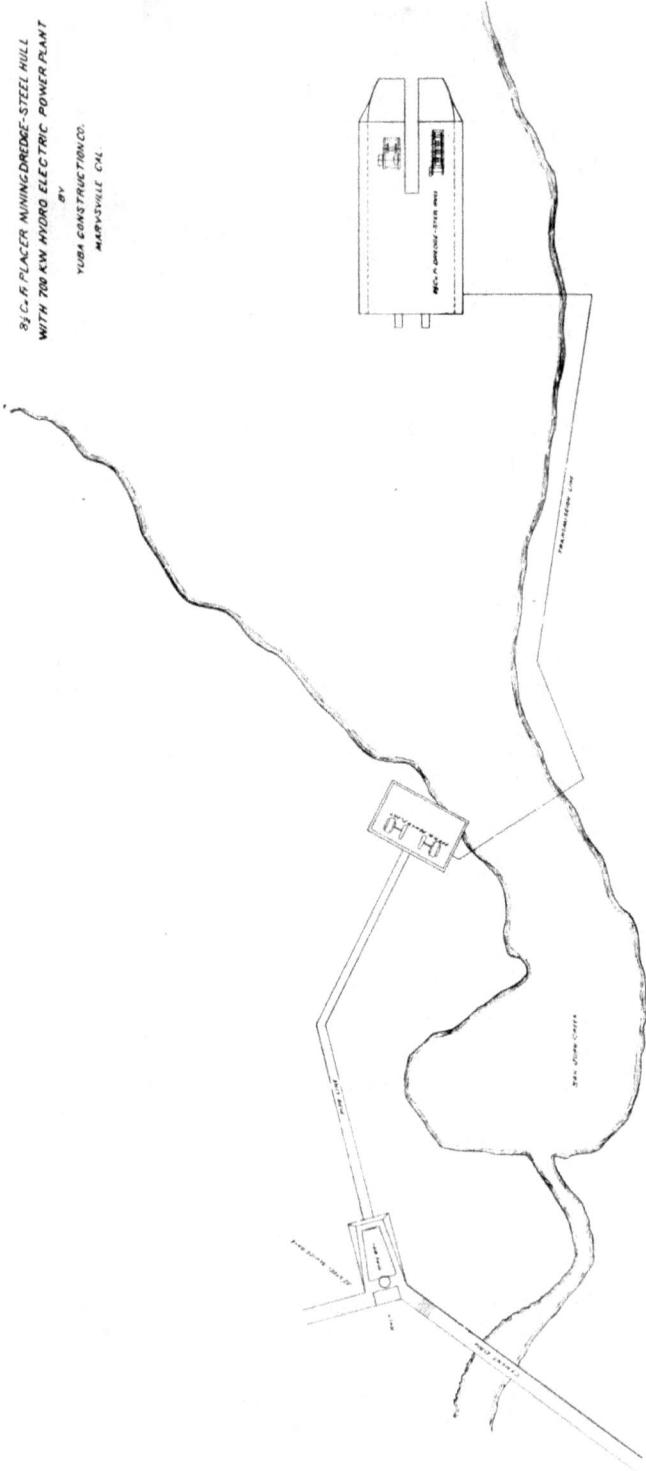

8½ C. F. PLACER MINING DREDGE-STEEL HULL
WITH 700 K.W HYDRO ELECTRIC POWER PLANT
BY
YUBA CONSTRUCTION CO.
MARYSVILLE CAL.

No. 238.   Plant of the Oroville Dredging, Limited, in Colombia, South America.

climate is tropical in every sense of the word, while the heavy and violent rainstorms during the wet season—lasting nearly nine months of the year—turn the flat areas suitable for dredging into veritable swamps which breed the malaria infected mosquito by the million. To guard against these pests requires a strong constitution and a sane manner of living which the northern man is loathe to adopt until he has gone through the mill.

A tropical jungle that is exceedingly dense and a cause of much annoyance must be cleared off most of the land. It must be understood that this country is very sparsely populated, and practically the only means of travel is by the large navigable rivers, or over frightfully rough and uncertain trails blazed through long stretches of impenetrable wilds. Small settlements exist at intervals along the rivers, but behind them there is always the jungle until another river is reached. As a consequence the traveler must carry his own camp equipment and food supplies upon any journey he intends to make into the interior and a plentiful supply of netting to keep away the insects at night.

The gold content of the ground is often high enough to prove very attractive for dredge mining, especially when it is taken into consideration that you can work throughout the year with no fear of an interruption due to cold weather. The only thing one must guard against is the chance of a heavy flood, if the dredge is working on a river. Such an occurrence is liable to happen at any time during the wet season, and occasionally during the dry season, with great rapidity and produce startling results. For this reason it is preferable to work inland a short distance where it is possible.

The gravel usually contains a small percentage of clay, but in most places this is not so much of a hindrance as the lack of uniform values. It is the usual thing for the owners to claim phenomenal values throughout their land, which must always be taken with a grain of salt. Rich spots are quite common, but their extent is usually rather limited, and one must discount these statements when a large area is figured on. There are so many rivers in this region flowing down from the upper Andes that sufficient water power can be developed in numerous places for all the dredges that are ever likely to be installed, although one may have to go some distance from his ground to find the most suitable location for a power house.

The conditions under which the men must live are very trying in some respects. A northerner going there for the first time should exercise the same precautions he would take if he lived in a swampy region here at home during very warm weather. All houses should be built off the ground upon some hilltop, have high ceilings with a double roof, and wide verandas. By clearing away the jungle from about the house, the warm sun will dry things up and a chance will be given to any breeze that may spring up. The soil over these low lands is exceedingly rich and fertile so that a large variety of vegetables can be raised at your very door, while cattle seem to thrive. To prepare the food, an imported cook is a necessity, as the native knows nothing about this important art. To keep good men contented in such a country nothing pays better than to set an attractive table.

The government of Colombia is now considered fairly stable and it is anxious to help all legitimate mining enterprises as much as possible. There is an export tax of 1 per cent upon all gold exported from the country; but mining machinery is practically free from duty. As far as labor goes, it is fairly efficient, for the wages paid. The peon is much like a small child and will work to the best of his ability, if treated properly. Probably the most difficult problem to solve for a foreign company, operating in this field, will be to secure the services of an efficient and honest corps of our countrymen to keep things going. It is a long way from home and the average man is apt to get homesick, if not really sick, and then his principal thought is how to get north with the least possible delay. This condition of things necessitates a reserve corps to fall back on when the occasion arises. In spite of the drawbacks, I look to see considerable capital invested in dredging enterprises in Colombia within the next few years.

DREDGING IN FRENCH GUIANA.

Until 1905 most of the alluvial gold had been obtained by crude methods of sluicing, but in that year a small dredge was put in commission on the Courcibo River, a tributary of the Sinnamari. For some time it gave good returns, but was eventually sunk during a flood. Two dredges were operating on the Lezard River in 1907.

Official figures give but a poor idea of the output of French Guiana. The Carsewene mine, which according to the owners produced gold to the value of 80,000,000 to 100,000,000 francs from the year 1894 to 1900, only supplied 21,000,000 to 22,000,000 francs worth to Cayenne. The Inini mine, on the Dutch frontier, shipped more than half the gold to Holland. Finally, many miners from the British Antilles carry away their own gold. All this gold was produced with the little Guiana sluice. Hitherto the small rivers or creeks of French Guiana have been mined chiefly by hand labor and small sluices, long toms, where there is not too much water nor too great a thickness of overburden. The overburden is often 2 to 3 metres deep, whereas the gold wash is only 30 to 40 centimetres. Gravel sand with 3 to 4 grammes of gold per cubic metre has been profitably worked. This seems a fair percentage, but circumstances are unfavorable; provisions are very dear,

and the alluvials very clayey and adhesive; consequently, dredges have
been adopted. At the end of the current year there will be five dredges
working in the colony, the type of which was only selected after many
experiments. The most conclusive were made at the Elyse placer, in
the River Lezard, and others in the Courcibo and Sparwin. The Crique
Roche has been carefully prospected. A dredge is being built for each
of these rivers. There are other rivers equally rich, yet to be opened
out. Results already obtained are as follows:

1. There is no difficulty in removing the wood buried in the alluvions
any more than large boulders; it is a matter of practice.

2. Washing is operated with jets of water in the trommel of the
dredge or (particularly when there is hard clay adhering to the
buckets) with picks. An automatic washer is being tried.

3. It has been found possible to instruct the natives in the management of a dredge. This is important, because, during the fifty years
gold has been known to exist, the labor problem has paralyzed all indus-
trial enterprise. There must, however, be a white overseer.*

The gold placers are mostly rich. Over stretches of 1 to 2 kilometres,
the average yield was 3 to 4 francs per cubic metre for a width of 30
to 40 metres. Dredging only exceptionally gave more, though this is
probably due to defective prospecting. Costs are still very great—1.80
to 2 francs per cubic metre. In any case, the alluvials, often very rich,
have been ample compensation for the absence of vein formations, and
likely to continue to be so, thanks to systematic dredging.

†Gold dredging in French Guiana seems to be successful when
working in the beds of the rivers. In view of this two modern gold
dredges are soon to be shipped to French Guiana to work on the deposits of the Sparwin River, a tributary of the Maroni (Ste. Anonyme
du Sparwin) and of the Courcibo River, a tributary of the Sinnamary
(Cie. Coloniale de Dragages Aurifères). They are of the open-link
continuous bucket type, with 4.5-cubic-foot buckets designed to work
at a depth of 9 metres. The pontoon and gantries are entirely made
of steel. A special feature of these dredges is a double revolving screen,
33 feet long, intended to disintegrate thoroughly the clayey alluvials
which are met with in Guiana. Besides the ordinary spray pipe leading
from the centrifugal pump, two rows of monitors, fed by a Knowles
pressure pump, will help to disintegrate the gravel. To prevent clay
from sticking on the Turner gold-saving tables, a special water jet
has been arranged over the tables.

†The Compagnie Coloniale de Dragages Aurifères and the Société
Anonyme du Sparwin report that the 5-cubic-foot gold dredges they
sent to French Guiana have arrived and are now under erection.

---

*The Mining Journal, from "La Nature."
†Engineering and Mining Journal.

DREDGING IN DUTCH GUIANA.

Considerable placer mining is being carried on in Dutch Guiana along the Saramaca, Surinamne, and Marowijne rivers and their tributaries. Like French Guiana, Dutch Guiana is very rich in alluvials, and most of the gold is produced by washing the gold-bearing material through sluices, or long toms, where there is not too much water or a too heavy overburden. Dredging for gold with small Holland-built dredges has been carried on along the Saramaca and Marowijne rivers. The climate is not bad and the country is rich in gold placers. Probably the best informed American on the gold placers of Dutch Guiana is Dr. W. H. Bradley, American Consul at Paramaribo, the capital town of Dutch Guiana.

BRITISH GUIANA.

The alluvial gold of British Guiana is found in the gravels of the existing streams. Only one hydraulicing company has carried on operations. This company, the Demerara Exploration Company, commenced work at Omai on the Essequibo River in 1902, but by 1907 had exhausted its sluicing ground and ceased operations in that direction. In 1904, the company placed a small dredge on Gilt Creek, and, its operations being successful, a larger dredge was placed in commission in June, 1906. Dredging at Omai meets with some difficulty on account of buried logs and trees. On the Conawaruk River a large dredge was erected and commenced work in January, 1907, but information is not available at the present writing as to the outcome.

*In 1907, about 80 per cent of the gold yield of British Guiana was produced by the individual miner and small parties. The total output being estimated at over $1,500,000.

GOLD DREDGING GROUND IN PERU AND BOLIVIA.

Alluvial gold occurs in several of the rivers of southern Bolivia. A recent attempt to dredge the gravels of the Rio San Juan de Oro, near Tupiza, resulted in a failure. One of the three dredges that had been in operation there was dismantled in 1908, and reconstructed on the Quebrada de Esmorca, a neighboring tributary of the San Juan.

The Tipuani River in the Larecja Province, is the richest in Bolivia. The auriferous gravels of the Tipuani are of great depth and true bedrock is seldom reached. Concentration of gold generally occurs on false bottoms of ferruginous conglomerates locally known as cargalli. Other placer deposits of value are at Yani, Tacacoma, and Chuquiaguillo.

*Gold, its Occurrence and Distribution, J. Malcolm Maclaren.

Placer deposits are numerous in Peru in the province of Sandia, north of Titicaca and southeast of Carabaza. The value of the gravels near Aporona has been estimated at 20 cents per cubic yard. Hydraulicking has been carried on with success in some districts, and the natives recover considerable gold by paving the dry beds of the streams with stones; the gold settling in the interstices between the stones during the floods of the wet season. The conglomerate gravels of the Poto are said to be from 60 to 180 feet thick.

Dredging on a large scale was being inaugurated in 1907 on the Rio Inambari in the Carabaya Province, and on the boundary between Peru and Bolivia. The placers of Pataz and Sandia, as well as those of the Rio Nusemescato, were also being investigated.

*Several American and English companies have been operating gold placer properties in Peru and Bolivia, at the headwaters of the Amazon. Roads have been built at large expense and considerable equipment taken in to the properties. The following description of the country and future possibilities is of interest.

Since the times of the Incas, rough trails have existed between the Inca centers of Titicaca and Cuzxo and the ravines of the eastern Cordilleras, which culminate in the Kaka River close to the small village of Guanay. The most important of these precipitous valleys have been formed on the rivers Tepuani, Mapiri, Challana, and Corvico.

A great amount of work was done in early days but nothing has been done in modern times save by an occasional prospector. The upper rivers are difficult to work because of the great boulders which are frequent and the torrential nature of the streams. They come together to form the Kaka just below Guanay and run through a succession of narrow canyons for about 20 miles, until Incahuara is reached. Below Incahuara the canyons of the Kaka wander out into great flats, often several miles wide, of shallow depth, and it is here that gravel deposits were formed.

Transportation is a serious difficulty. Rough mule trails are the only means of entering this region and the transportation of heavy machinery is almost impossible until roads of some kind are built.

†Practically all that has been done in prospecting the Incahuara and Inambari gravels has been to ascertain surface richness. At Incahuara, however, the ground was found to contain value, the total depth explored, of 16 feet. At no place has bedrock been reached, neither has the bed of either river been examined. Companies are now proceeding to test these gravels by actual working.

The only practical method of working these gravels is dredging.

---

*Abstract from report of Alexander Benison, La Paz, in the Engineering and Mining Journal.

†Abstract from report of M. Conway.

The ground does not fall steeply enough for sluicing or hydraulicking. At Incahuara, for instance, and along most of the reaches of the lower Inambari (except where torrential side streams enter it) there are no boulders, and on neither river is there hard clay; the gravel is composed of loose stones and sand.

At Incahuara the bedrock is sandstone and can be easily scraped by the lip of the dredge buckets; a similar bedrock appears on the Inambari. The current of neither river is swift enough to cause danger to the dredge.

The transportation of dredges to these remote places is difficult and costly; in the case of the Incahuara dredge, a considerable part of the machinery is said to have been already transported to the site. According to report, the Ollachea road presents no greater difficulties than does the Caravani.

The only part of the ground on the Inambari and Beni rivers that has been examined by competent men is the Incahuara basin, which contains about 10,000,000 cubic yards of dredgeable gravel. The surface has nowhere yielded less than 19 cents a cubic yard, while large areas average at least a dollar. An average value of 75 cents per cubic yard is not an extravagant estimate; of course, local authorities put the average much higher. The gold from the Inambari resembles, in shape and quality, the gold obtained from the Kaka or Beni River. It is flat and relatively thick.

For 100 miles below the Incahuara basin it is believed that the Beni gravels yielded gold. There are numbers of places that have been definitely located, so that in all probability the Incahuara basin is a mere fractional part of the dredgeable ground on the Beni.

On the Inambari, if no spot has been so carefully examined as the Incahuara, a much longer stretch of the river has been rapidly sampled by more than one prospector, high values to the cubic yard being revealed. The Inambari, therefore, is likely to yield as much gold as the Beni.

*GOLD PLACERS OF TIERRA DEL FUEGO.

Until recently the largest gold mining operations were at Paramo and Lennox Island, but since the introduction of dredges the most active operations are on the northwestern part of the main island of Tierra del Fuego, across the strait of Magellan from Punta Arenas. Here the town of Powenir is the headquarters of the industry.

This town is now a prosperous mining center of about 800 people. In addition to the Powenir region mining on a small scale of more or less importance is still going on at some of the other localities.

*Abstracts from papers by R. A. Penrose, Jr., S. H. Loram, and from "Estadistica Minera de Chile" of 1908.

The chief center of civilization in the whole region is the Chilean town of Punta Arenas, on the Patagonian side of the strait. This town has a fairly good harbor, is an active place of 12,000 population and the seat of government for this part of the Chilean possessions. The settlement of Ushuwaia, the seat of government for the Argentine part of Tierra del Fuego, is only a small place. Punta Arenas is in 50° 9′ 42″ south latitude, and has the distinction of being the most southerly town of any considerable size in the southern hemisphere. Those interested in the geology of the region will find useful information in "Geological Observations in South America" by Charles Darwin.

In Tierra del Fuego the surface is rolling country, cut through by seaworn valleys and straits, and terminating on the east coast in what is really a vertical section along that particular line, with some cliffs 50 to 60 metres high. Judging by the pebbles the deposit was derived from the denudation of hornblende-quartz-schist, diorite, syenite, andesite, and clay-slates. The first three of these are well known, as the country rocks contain gold-bearing lodes, besides carrying small quantities of the precious metal themselves, consequently the detritus from them might be expected also to contain gold, even if in minute quantities, as in the present case. Once above sea level denudation started on the accumulated soft beds by wave action; wide bays, inlets and channels were cut, the lighter particles being carried away, while the gravel and heavier constituents, as oxide of iron, garnet, gold, and the like, remained behind in concentrated form.

Nearly all the gold of the Magellan region, so far as known, is in alluvial deposits; few gold-bearing veins have been found. The alluvial deposits may be divided into two classes, those in beds of creeks or hillsides, and those on sea beaches, where they are subject to the action of the sea; during the high spring tide, the rise and fall of the tide is 45 feet and in winter there are occasional furious storms. This action of the sea has had much to do with the forming of the richer deposits on the beaches. Mining by hand labor and washing the beach sand and gravel in long-toms has been carried on with more or less success for some years. In the mineral statistics of Chile it is estimated that the production for the whole region during 1903 and 1904 was 137 and 170 kilograms, respectively, practically all of which came from the small workings.

The alluvial deposits in beds of streams and in hillsides vary in content from a few cents to $1 or more a cubic yard. Most of the ground now worked is claimed to average from 25 to 50 cents per cubic yard. Under the existing conditions it is difficult to make low grade ground pay. The gold-bearing beds vary from a few feet to 10 or 30 feet deep; an overburden of barren ground often occurs.

Prospecting is a more difficult task than in most places. Traveling is done mostly in boats, as the land is cut up by deep tide water channels and bays and covered with dense underbrush or immense peat bogs, while everywhere, even on the mountain sides, the soil is soft and boggy, so that walking is difficult and often impossible.

The climate, though stormy, is not extreme, the thermometer rarely going much below zero, or greatly above 60° Fahrenheit; the mean temperature in winter being about 33° and in summer about 50° Fahrenheit. The season during which mining can be carried on is from August to May.

The history of the dredging operations to the present time has, with few exceptions, been a succession of failures, caused, on the main, by ill-advised projects or wild-cat promoters, and to a smaller extent by the use of badly designed and poorly constructed dredges.

The first company to install a dredge, according to the "Estadistica Minera de Chile," was the Compania Sutphen de Lavaderos de Oro, which was formed in Buenos Aires in 1903, but on account of a series of delays from various causes £30,000 and two years' time were spent before the dredge was ready to begin operations on the Rio del Oro. According to S. H. Loram this dredge was built to handle small gravel only, and was not provided with grizzlies or trommel, whereas operations showed up many boulders of various sizes, some large enough to stall the digging engine when encountered. Breakdowns were frequent on account of the clay bottom; considerable difficulty was experienced in washing the gravel and in getting the buckets to dump. However, results were sufficiently encouraging to permit an increase in the capitalization of the company and four large dredges were ordered from Holland.

Meanwhile, the gold excitement had spread and land was taken up all over the island. Glowing reports were issued ofttimes without the slightest knowledge of conditions. One company in the most accessible part of the island began work on the basis of two drill holes, one at either end of the property. The majority of the companies did not even go to that trouble until after the dredge was installed on the property.

A significant feature was that gold dust and nuggets were at a premium of nearly 100 per cent. Generally the showing of dust supposed to have come from a property backed by the statements of an "engineer" (who was to receive the greater part of his pay in shares of the company, if formed) was sufficient to find all the capital called for. Below is given an incomplete list of the companies formed, with the capitalization of each. To these must be added a considerable amount of private capital and also those companies that did not get organized in time for the boom. The shares of most of these listed companies were at one time quoted at a premium.

*List of Principal Companies Formed to Work Gold in Magellan Territory 1903-1906.

Cia. Sutphen de Lavaderos de Oro, later increased to	†2,000,000
Cia. Dragaje Rio del Oro de Tierra del Fuego	‡35,000
Cia. Dragaje Rio Verde	‡70,000
Soc. Lavaderos de Oro de Tierra del Fuego	‡300,000
Cia. Dragaje de Rio Gallego Chico	‡35,000
Cia. Dragaje de Rio Palo	‡35,000
Cia. Dragaje de Rio San Martin	‡62,000
Cia. Dragaje de America	‡100,000
Cia. Rios Unidos de Tierra del Fuego	‡40,000
Cia. Dragaje del Rio Progreso	‡70,000
Cia. Exploradora de Rio Grande	‡40,000
Soc. Exploradora de Ultima Esperanza	‡10,000
Soc. Chorillos de Rio del Oro	‡50,000
Rio Oscar Dredging Co.	†250,000
Soc. Exploradora de Minas de Magallanes	‡15,000
Soc. Carmen Silva	‡35,000
Cia. Argentina de Esploracion en Tierra del Fuego	‡400,000
Cia. Aurifera de Punta Delgada	†750,000
Cia. Minera Rio Colorado	†300,000
Cia. Loreto	
Cia. Aurifera de Lennox	
Cia. Brunswick	
Cia. Rosario	

Of the above list, according to Mr. Loram, the only companies carrying on active operations to-day are the Sutphen and Lennox, with what results is unknown. Of the rest it is difficult to get particulars, and their shares are no longer quoted. Three, the Rio del Oro, Rio Verde, and Rio Progreso, which were formed, floated and managed by John D. Roberts, once well digger, then dredge expert, and transitory Gold King of Sandy Point, liquidated in June, 1908. The first two lost all their capital and had some small debts, the third was more fortunate and expected to return 33 pence for each £1 share.

According to a statement issued to the stockholders the dredge started work November 11, 1907, and stopped March 14, 1908, had run 80 days and been stopped 44. During this time it produced 8,988 grammes gold (fineness not stated) corresponding to 112 grammes per working day. The secretary of the company stated that he believed the capacity of the dredge to be 1,000 cubic metres per day, so the production would work out 0.112 grammes per cubic metre, or .0082 of an ounce per cubic yard.

According to the Estadistica Miner most of the dredges seem to have been constructed in Holland; they were operated with steam power, and coal has been the fuel most used. On account of transportation the cost of coal is excessive and experiments have been made to burn turf or peat with, it is stated, good results.

Most of the dredging companies, so far as known, were of Argentine or Chilean origin. The Queen Gold Dredging Company of London acquired land on Tierra del Fuego in 1908 and installed a dredge

*By S. H. Loram.   †Argentine pesos.   ‡Chilean pesos.

designed by Cutten Bros. In July, 1909, however, according to The Mining Manual, it was resolved to wind up the affairs of the company voluntarily; data as to actual operations, if any, is not furnished.

GOLD DREDGING IN ECUADOR.

Little information is available of the mineral resources of Ecuador. The United States Gold Dredging and Rubber Company of New York is interested in a placer deposit in the province of Esmeralda, and Coats & Associates of London also control large property interests in the same province. The Ecuadoran government is anxious to interest foreign capital to develop mining enterprises, and has established liberal mining laws.

GOLD DREDGING IN BRAZIL.

Of late years dredging companies have been floated in Buenos Aires to dredge rivers descending from the Matto Grosso central hills. They have almost all been failures, however, although gold undoubtedly exists. The difficulty of transportation is a serious drawback. The most accessible parts of Matto Grosso are a three weeks' journey up the Parana River system, and communication is unreliable.

A dredge was installed in 1902, to work the gravels of the Coxipo de Ouro, which flows into the Cuyaba. These gravels had been worked for many years by native methods. It is claimed that this dredge operated successfully. Other dredging areas are on the Piracicaba River, which flows into the Rio Tiete, a tributary of the Parana. In the province of Minas Gereas, dredges are in operation on the Rio das Mortes and the Ribierao de Carmo.

*The following is given as a list of operating Rio Plata dredging companies: Brumado Gold Dredging Company, Cabaeal Gold Dredging Company, Matto Grosso Gold Dredging Company, Diamantino (Matto Gross), Este Matto Grosso Company.

GOLD DREDGING IN ARGENTINA.

Auriferous gravels are found in the Jujuy province, also at Famatina, and in the eastern portion of Tierra del Fuego. Gold is also found in the provinces of San Luis, San Juan, Tacuman, Calamarca and Salta.

There have been some minor dredging operations in the northern rivers which, however, do not seem to have been very successful, though several dredges are reported to be still working. At the head of the Neuquen River, in the Cordilleras, the auriferous area is said to be of considerable extent and has been successfully worked for years by Chileans, in a small way by ground sluicing.

*Bulletin of the Bureau of American Republics.

*DREDGING IN KELANTAN, MALAY PENINSULA.

The Duff Development Company own the only four dredges on the Kelantan River, which flows to the east coast of the Malay Peninsula, and which is 600 to 800 feet wide, about 100 miles inland, at the place where the dredges have been working. During two months one of the dredges won 1,600 ounces of gold. The biggest return the company has had from one dredge was 375 ounces for one week. Each dredge has a dredgemaster, three European winchmen—one for each shift—and about 16 native hands. The dredgemasters received a salary of about $2,000, the European winchmen about $1,500, and the native coolies 50 cents per day. Firemen and greasers got 60 cents and native winchmen 70 cents. Expenses are considerably heavier than in New Zealand, but a man could have everything necessary for about $10 per week. Generally speaking, the climate is good. It is moist and warm, but not so oppressive as is generally supposed. The rainy season usually lasts only a week or a fortnight, and at such times the rivers rise perhaps 30 or 40 feet in 24 hours, but fall again just as quickly. There are numbers of other smaller rivers on the Malay Peninsula on which nothing has ever been done beyond a little prospecting.

KOREAN GOLD PLACERS.

†Placer gold is found in almost every prefecture in Korea; but practically all the deposits are too low grade to attract foreign capital and the richer ones have mostly been worked out. The native method of mining is by sluicing; a small stream of water is conducted through a ditch lined with stones. The gravel is excavated with picks and the fine gravel shoveled into the ditch, the boulders being thrown to one side. The gold is recovered by panning the concentrates removed from the sluice in large wooden pans. No mercury is used.

The Koreans as laborers are inferior to the Chinese, except perhaps as underground miners, for which they seem to have a natural aptitude. The average Shantung coolie is strong, energetic, good natured, and usually tractable. The best feature of the labor situation in Korea is the fact that Chinese and Koreans do not pull together and a strike by one nationality may always be broken by employing the other. The climate is healthful, water is abundant, and power may frequently be developed at small expense. Transportation is by bullock cart or pack horses. Fuel and timber are scarce, which is a serious factor in successful operations of mining enterprises. Where timber is found, it is usually oak or pine. The only extensive forest areas are in the extreme north of the country, near the Manchurian frontier.

---

*From the Mining and Scientific Press, 1909.
†Abstract from "The Mineral Resources of Korea," by H. R. Robins, Trans. A. I. M. E., Vol. 39.

*The District of Chiksan has been one of the most profitable placer fields worked by the Koreans. For some time past Japanese have been successfully working portions of these placers by sluicing. The conditions for successful work of this kind have been unsatisfactory. There being comparatively little water and a very slight fall, all the pay dirt has been carried to the sluice boxes on the backs of Korean coolies.

The pay gravel occurs principally in a stratum of about 3 feet on bedrock. Recent investigations have proved that there are values in the overburden, though not in sufficient quantity to make it profitable to run the same through the sluices. The depth of the overburden varies from 7 to 30 feet, most of it being about 18 feet in depth. There are no boulders larger than 8 or 10 inches, the greater portion of the overburden being gravel with stones about the size of an egg. The bedrock is soft decomposed schist, which may be dug with a spade to a depth of 2 or 3 feet. While there is very little water on the surface, there would be sufficient for dredge operation.

So far as can be determined from available information, there have been no gold dredges operated in Korea. Tests have been made of different placer areas, but apparently with indifferent results; according to the Consular Report for 1907, tests in the Usan fields demonstrated average values from 5 to 6 cents only. It has been recently reported that some Japanese companies were to install dredges and had sent representatives to California to gain information of operating methods and dredge construction.

DREDGING IN WEST AFRICA.

In the Gold Coast Colony where dredging operations have been carried on for some years, the principal dredging rivers are the Offin Ankobra and Birrim.

In French Guinea dredges have been working on the Tankisso River and auriferous areas are being examined in the district between the Senegal and Niger rivers where dredges will be installed if the engineers report favorably.

---

*Abstract from United States Consular Report for 1907.

Ferry Building, San Francisco, one half the upper floor of which is occupied by the State Mining Bureau. (This building is constructed of Colusa sandstone, and the reconstructed tower is of reinforced concrete.)

APPENDIX.

CALIFORNIA STATE MINING BUREAU.

This institution is the chief source of reliable information about the mineral resources and mining industries of California.

It is encouraged in its work by the fact that its publications have been in such demand that large editions are soon exhausted. In fact, copies of them now command high prices in the market.

The publications, as soon as issued, find their way to the scientific, public, and private libraries of all countries.

STATE MINERALOGIST.

The California State Mining Bureau is under the supervision of a State Mineralogist and Board of Trustees.

It is supported by legislative appropriations, and in some degree performs work similar to that of the geological surveys of other states, but its purposes and functions are mainly practical, the scientific work being clearly subordinate to the economic phases of the mineral field, as shown by the organic law governing the Bureau, which is as follows:

SEC. 4. It shall be the duty of said State Mineralogist to make, facilitate, and encourage special studies of the mineral resources and mineral industries of the State. It shall be his duty: To collect statistics concerning the occurrence of the economically important minerals and the methods pursued in making their valuable constituents available for commercial use; to make a collection of typical geological and mineralogical specimens, especially those of economic or commercial importance, such collection constituting the Museum of the State Mining Bureau; to provide a library of books, reports, drawings, bearing upon the mineral industries, the sciences of mineralogy and geology and the arts of mining and metallurgy, such library constituting the Library of the State Mining Bureau; to make a collection of models, drawings, and descriptions of the mechanical appliances used in mining and metallurgical processes; to preserve and so maintain such collections and library as to make them available for reference and examination, and open to public inspection at reasonable hours; to maintain, in effect, a bureau of information concerning the mineral industries of this State, to consist of such collections and library, and to arrange, classify, catalogue, and index the data therein contained, in a manner to make the information available to those desiring it, and to provide a custodian specially qualified to promote this purpose; to make a biennial report to the Board of Trustees of the Mining Bureau, setting forth the important results of his work, and to issue from time to time such bulletins as he may deem advisable concerning the statistics and technology of the mineral industries of this State.

THE BULLETINS.

The field covered by the books issued under this title is shown in the list of publications. Each bulletin deals with only one phase of mining. Many of them are elaborately illustrated with engravings and maps. Only a nominal price is asked, in order that those who need them most may obtain a copy. (See list on last page.)

THE REGISTERS OF MINES.

The Registers of Mines form practically both a State and a County directory of the mines of California, each county being represented in a separate pamphlet. Those who wish to learn the essential facts about any particular mine are referred to them. The facts and figures are given in tabular form, and are accompanied by a topographical map of the county on a large scale, showing location of each mineral deposit, towns, railroads, roads, power lines, ditches, etc.

HOME OF THE BUREAU.

The Mining Bureau occupies the north half of the third floor of the Ferry Building, in San Francisco. On the same floor are the rooms of the California Development Board, and an exhibition instituted by and maintained by the same Board, illustrative of agriculture, horticulture, viticulture, and other industries of California. All visitors and residents are invited to inspect the Museum, Library, and other rooms of the Bureau and gain a personal knowledge of its operations.

THE MUSEUM.

The Museum now contains over 20,000 specimens, carefully labeled and attractively arranged in showcases in a great, well-lighted hall, where they can be easily studied. The collection of ores from California mines is of course very extensive, and is supplemented by many cases of characteristic ores from the principal mining districts of the world. The educational value of the exhibit is constantly increased by substituting the best specimens obtainable for those of less value.

These mineral collections are not only interesting, beautiful, and in every way attractive to the sightseers of all classes, but are also educational. They show to manufacturers, miners, capitalists, and others the character and quality of the economic minerals of the State, and where they are found. Plans have been formulated to extend the usefulness of the exhibit by special collections, such as one showing the chemical composition of minerals; another showing the mineralogical composition of the sedimentary, metamorphic, and igneous rocks of the State; the petroleum-bearing formations, ore bodies, and their country rocks, etc. A fine permanent exhibition of the structural materials of California will be maintained to illustrate the resources of the State in that regard.

Besides the mineral specimens, there are many models, maps, photographs, and diagrams illustrating the modern practice of mining, milling, and concentrating, and the technology of the mineral industries. An educational series of specimens for high schools has been inaugurated, and new plans are being formulated that will make the Museum even more useful in the future than in the past. Its popularity is shown by the fact that more than 125,000 visitors registered last year, while many failed to leave any record of their visit.

THE LIBRARY.

This is the mining reference library of the State, constantly consulted by mining men, and contains about 5,000 volumes of selected works, in addition to the numerous publications of the Bureau itself. On its shelves will be found reports on geology, mineralogy, mining, etc., published by states, governments, and individuals; the reports of scientific societies at home and abroad; encyclopædias, scientific papers, and magazines; mining publications, and the current literature of mining ever needed in a reference library.

Manufacturers' catalogues of mining and milling machinery by California firms are kept on file. The Registers of Mines form an up-to-date directory for investor and manufacturer.

The Librarian's desk is the general bureau of information, where visitors from all parts of the world are ever seeking information about all parts of California.

READING-ROOM.

This is a part of the Library Department and is supplied with more than one hundred current publications. Visitors will find here various California papers and leading mining journals from all parts of the world.

The Library and Reading-Room are open to the public from 9 A. M. to 5 P. M. daily, except Sundays and holidays.

THE LABORATORY.

This department identifies for the prospector the minerals he finds, and tells him the nature of the wall rocks or dikes he may encounter in his workings; but this department *does not* do assaying nor compete with private assayers. The presence of minerals is determined, but not the percentage present. No charges for this service are made to any resident of the State. Many of the inquiries made of this department have brought capital to the development of new districts. Many technical questions have been asked and answered as to the best chemical and mechanical processes of handling ores and raw material. The laboratory is well equipped.

THE DRAUGHTING-ROOM.

In this room are prepared scores of maps, from the small ones filling only a part of a page, to the largest County and State maps; and the numerous illustrations, other than photographs, that are constantly being required for the Bulletins and Registers of Mines. In this room, also, will be found a very complete collection of maps of all kinds relating to the industries of the State, and one of the important duties of the department is to make such additions and corrections as will keep the maps up to date. The seeker after information inquires here if he wishes to know about the geology or topography of any district; about the locations of the new camps, or positions of old or abandoned ones; about railroads, stage roads, and trails; or about the working drawings of anything connected with mining.

MINERAL STATISTICS.

One of the features of this institution is its mineral statistics. Their annual compilation by the State Mining Bureau began in 1894. No other state in the Union attempts so elaborate a record, expends so much labor and money on its compilation, or secures so accurate a one.

The State Mining Bureau keeps a careful, up-to-date, and reliable but confidential register of every producing mine, mine-owner, and mineral industry in the State. From such are secured, under pledge of secrecy, reports of output, etc., and all other available sources of information are used in checking, verifying, and supplementing the information so gained. This information is published in an annual tabulated, statistical, single-sheet bulletin, showing the mineral production by both substances and counties.

TOTAL GOLD PRODUCT OF CALIFORNIA—1848-1908.

While gold is next to the leading mining product, its yield no longer puts the greatest gold-producing county in the first place. Gold is more widely distributed than any other substance thus far mined in California.

The following table shows the total gold yield of California, by years, from the time mining commenced in 1848 to 1908, inclusive:*

Year	Amount	Year	Amount
1848	$245,301	1880	$20,030,761
1849	10,151,360	1881	19,223,155
1850	41,273,106	1882	17,146,416
1851	75,938,232	1883	24,316,873
1852	81,294,700	1884	13,600,000
1853	67,613,487	1885	12,661,044
1854	69,433,931	1886	14,716,506
1855	55,485,395	1887	13,588,614
1856	57,509,411	1888	12,750,000
1857	43,628,172	1889	11,212,913
1858	46,591,140	1890	12,309,793
1859	45,846,599	1891	12,728,869
1860	44,095,163	1892	12,571,900
1861	41,884,995	1893	12,422,811
1862	38,854,668	1894	13,923,281
1863	23,501,736	1895	15,334,317
1864	24,071,423	1896	17,181,562
1865	17,930,858	1897	15,871,401
1866	17,123,867	1898	15,906,478
1867	18,265,452	1899	15,336,031
1868	17,555,867	1900	15,863,355
1869	18,229,044	1901	16,989,044
1870	17,458,133	1902	16,910,320
1871	17,477,885	1903	16,471,264
1872	15,482,194	1904	19,109,600
1873	15,019,210	1905	19,197,043
1874	17,264,836	1906	18,732,452
1875	16,876,009	1907	16,727,928
1876	15,610,723	1908	18,761,559
1877	16,501,268	1909†	21,500,000
1878	18,839,141		
1879	19,626,654	Total	$1,509,775,250

* Figures for 1906, 1907, and 1908 by U. S. Geological Survey.   † Estimated.

LIST OF PUBLICATIONS.

Publications of this Bureau will be sent on receipt of the requisite amount and postage. *Only stamps, coin or money orders will be accepted in payment. Do not send personal checks.*

Address all communications regarding publications to LIBRARIAN.

(All publications not mentioned are exhausted.)

	Price.	Postage.
Report XI—1892, First Biennial	$1.00	$0.15
Report XIII—1896, Third Biennial	1.00	.20
Bulletin No. 6—"Gold Mill Practices in California" (3d ed.)	.50	.04
Bulletin No. 9—"Mine Drainage, Pumps, Etc." (bound)	.60	.08
Bulletin No. 15—"Map of Oil City Oil Fields, Fresno County, California"	.05	.02
Bulletin No. 23—"Copper Resources of California"	.50	.12
Bulletin No. 27—"Quicksilver Resources of California" (2d ed.)	.75	.14
Bulletin No. 30—"Bibliography Relating to the Geology, Palæontology and Mineral Resources of California," including List of Maps	.50	.10
Bulletin No. 31—"Chemical Analysis of California Petroleum"	---	.02
Bulletin No. 32—"Production and Use of California Petroleum"	.75	.08
Bulletin No. 36—"Gold Dredging in California" (3d ed.)	.50	.08
Bulletin No. 37—"Gems and Jewelers' Materials of California" (2d ed.)	.50	.08
Bulletin No. 38—"Structural and Industrial Materials of California"	.75	.20
Bulletin No. 45—"Auriferous Black Sands of California"	.10	.02
Bulletin No. 46—"Index of Mining Bureau Publications"	.30	.06
Bulletin No. 50—"Copper Resources of California" (revised ed.)	1.00	.20
Bulletin No. 57—"Gold Dredging in California" (revised ed.)	---	---
Bulletin No. 58—"Mineral Production of California"—1909	---	.02
Bulletin No. 59—"Mineral Production of California for 23 Years"	---	.02
Bulletin No. 60—"Minerals of California, Mining Laws, Maps, Etc."	---	.05
California Mine Bell Signals (cardboard)	.05	.02
California Mine Bell Signals (paper)	.03	.02
Register of Mines, with Map, Amador County	.25	.08
Register of Mines, with Map, Butte County	.25	.08
Register of Mines, with Map, El Dorado County	.25	.08
Register of Mines, with Map, Inyo County	.25	.08
Register of Mines, with Map, Kern County	.25	.08
Register of Mines, with Map, Lake County	.25	.08
Register of Mines, with Map, Mariposa County	.25	.08
Register of Mines, with Map, Nevada County	.25	.08
Register of Mines, with Map, San Bernardino County	.25	.08
Register of Mines, with Map, San Diego County	.25	.08
Register of Mines, with Map, Santa Barbara County	.25	.08
Register of Mines, with Map, Shasta County	.25	.08
Register of Mines, with Map, Sierra County	.25	.08
Register of Mines, with Map, Siskiyou County	.25	.08
Register of Mines, with Map, Trinity County	.25	.08
Register of Mines, with Map, Tuolumne County	.25	.08
Register of Mines, with Map, Yuba County	.25	.08
Register of Oil Wells, with Map, Los Angeles City	.35	.02
Map of El Dorado County Showing Boundaries National Forests	.20	.02
Map of Madera County Showing Boundaries National Forests	.20	.02
Map of Placer County Showing Boundaries National Forests	.20	.02
Map of Shasta County Showing Boundaries National Forests	.20	.02
Map of Sierra County Showing Boundaries National Forests	.20	.02
Map of Siskiyou County Showing Boundaries National Forests	.20	.02
Map of Trinity County Showing Boundaries National Forests	.45	.02
Map of Tuolumne County Showing Boundaries National Forests	.20	.02
Map of Mother Lode	.05	.02
Map of Desert Region of California	.10	.02
Map Showing Copper Deposits in California	.05	.02
Map of Calaveras County	.25	.03
Map of Placer County	.25	.03
Map of Plumas County	.25	.03
Mineral and Relief Map of California	.25	.05
Map of Forest Reserves in California (mounted)	.50	.08
Map of Forest Reserves in California (unmounted)	.30	.06
Map of Minaret District, Madera County	.20	.02
In Preparation—		
Bulletin—"Petroleum in California"	---	---

INDEX.

www.ingramcontent.com/pod-product-compliance
Lightning Source LLC
Chambersburg PA
CBHW060321200326
41519CB00011BA/1795